高校土建类专业英文规划教材

Seismic Design of Building Structures
建筑结构抗震设计

张　海　刘中宪　何　颖　王　岱 编著

中国建筑工业出版社

图书在版编目（CIP）数据

建筑结构抗震设计＝Seismic Design of Building Structures/
张海等编著.—北京：中国建筑工业出版社，2020.6（2022.6重印）
高校土建类专业英文规划教材
ISBN 978-7-112-24964-0

Ⅰ.①建… Ⅱ.①张… Ⅲ.①建筑结构-防震设计-
高等学校-教材-英文 Ⅳ.①TU352.104

中国版本图书馆 CIP 数据核字（2020）第 041571 号

This book is written by the current *Syllabus for China Undergraduate of Civil Engineering*, *The Code for Seismic Design of Buildings* GB 50011—2010 and other relevant China codes and technical specifications. It can be used as a teaching book for the course of "seismic design of building structures" in the majors related to civil engineering and a teaching reference for the course of "seismic design of building structures" for international students. This book has six chapters including basic knowledge of earthquake resistance, seismic concept design of building, site, base and foundation, dynamics of structures and seismic response, seismic design of reinforced concrete structures and earthquake resistance of underground structures. Some examples and exercises are included in each chapter.

This book is sponsored and published by Tianjian Science and Technology Association.

责任编辑：李笑然　吉万旺
责任校对：李美娜

高校土建类专业英文规划教材
Seismic Design of Building Structures
建筑结构抗震设计
张　海　刘中宪　何　颖　王　岱　编著

＊

中国建筑工业出版社出版、发行（北京海淀三里河路 9 号）
各地新华书店、建筑书店经销
北京鸿文瀚海文化传媒有限公司制版
北京建筑工业印刷厂印刷

＊

开本：787×1092 毫米　1/16　印张：10½　字数：262 千字
2020 年 7 月第一版　2022 年 6 月第二次印刷
定价：32.00 元
ISBN 978-7-112-24964-0
（35723）

Foreword

In recent years, with the increase of international exchange of education in China, more and more international students come to Chinese universities to study for degrees or join the joint training double-degree program. Based on the experience of joint training of Chinese and foreign university students for many years and the experience of seismic design of building structures for international students since 2009, the teaching book for *Seismic Design of Building Structures* is finished through the arrangement and summary of course teaching contents.

The book mainly introduces the outstanding achievements in earthquake resistance of ancient Chinese buildings and modern engineering structures, seismic calculation principle of building structures and seismic design method of underground structures. This book includes basic knowledge of earthquake resistance, seismic concept design of building, site, base and foundation, dynamics of structures and seismic response, seismic design of reinforced concrete structures and earthquake resistance of underground structures. The book focuses on two-stage design method, seismic concept design, response spectrum theory, internal forces and deformation of frame structure and calculation method of underground structure. Furthermore, some examples are given to help readers to understand the application of significant theories.

This book can be used as a teaching book for undergraduates in the majors related to civil engineering of colleges and universities, as a teaching book for international students, especially those of the countries along the Belt and Road (B&R), to make international students with different knowledge background master seismic theory of building structures and participate in the construction and practice of B&R project, and also as a reference for students and professionals who want to learn English vocabulary and its application about earthquake resistance of building structures, or to improve their interest and ability in learning English.

In the preparation of this book, we have referred to and quoted some publicly published literature and we would like to thank these authors. Due to the limited knowledge of editors, some omissions could appear in this book, please contact us if you have any suggestions.

This book is sponsored and published by Tianjian Science and Technology Association.

Directory

Chapter 1　Basic Knowledge of Earthquake Resistance

From time immemorial, the nature's forces have influenced human existence. Even in the face of catastrophic natural phenomena, human beings have tried to control nature and coexist with it. In the natural disasters, for example, earthquakes, floods, tornadoes, hurricanes, droughts, and volcanic eruptions, the least understood and the most destructive are earthquakes. Although the average annual losses due to floods, tornadoes, hurricanes, etc., exceed those due to earthquakes, the total, unexpected, and nearly instantaneous devastation caused by a major earthquake has a unique psychological impact on the affected. Thus, this significant life hazard demands serious attention. We need to focus on the hazards of earthquakes and how to resist the occurrence of earthquakes in this chapter. This chapter mainly introduces the causes of earthquakes, the related concepts of earthquakes, ground motions and so on. Let's learn together.

1.1　Causes of earthquakes

Earthquakes are natural phenomenon of tectonic movements within the Earth and may be defined as a wave-like motion generated by forces in constant turmoil under the surface layer of the earth, travelling through the earth's crust. The Earth has an average of 5 million earthquakes per year. Among them, if a strong earthquake affects human beings, it will cause an earthquake disaster and bring serious personal injury and economic losses to human beings. In order to alleviate the earthquake damage of buildings, avoid casualties, and reduce economic losses, engineers must carry out seismic analysis and seismic design of construction projects.

1.1.1　Earth internal structure

Earth's structure: equatorial radius $R=6371$km, polar radius 6357km, average radius about 6400km, including: crust, mantle and core (Figure 1.1).

The lithosphere is the hard rock of the upper part of the earth relative to the asthenosphere. It is about $60 \sim 120$ kilometers thick and is a high wave velocity belt for earthquakes. Including the entire crust and the top of the upper mantle, composed of granitic rocks, basaltic rocks and ultrabasic rocks.

The crust is composed of various uneven rocks and is the outermost layer of the Earth's solid circle. Except for the sedimentary layer on the ground, the crust below the

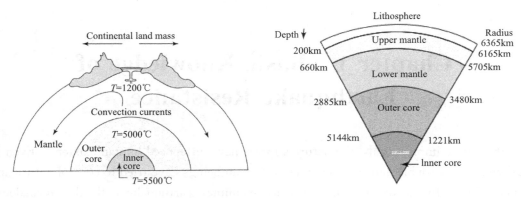

Figure1. 1   Internal structure of the Earth

land is mainly composed of a granite layer on the upper part and a basalt layer on the lower part. The crust below the ocean generally has only basalt layers, and the thickness varies from place to place. Most of the world's earthquakes occur in this thin crust.

The mantle is mainly composed of hard-coated peridotite, which has viscoelasticity. Since the radioactive material in the earth continuously releases energy, from the underground 20~700km, the internal temperature of the earth is about 600~2000℃, within this range. There is a soft layer in the mantle that is about a few hundred kilometers thick. The material convection and the pressure inside the earth are not balanced (900~370000MPa), the internal material of the mantle slowly moves under the action of heat and unbalanced pressure, which may be the roots of crustal movement.

The centrosphere is the core part of the earth. It is divided into the outer core (thickness 2100km) and the inner core. The main constituent materials are nickel and iron. It is speculated that the outer core may be in a liquid state and the inner core may be in a solid state.

1. 1. 2   Type of earthquake

Earthquakes cause disasters to human beings, such as different levels of casualties and economic losses. For example, in the 20th century, the number of deaths caused by earthquakes in the world in the first 80 years (1900-1980) was 1. 05 million, with an average of 13, 000 deaths per year. In 1990, more than 50, 000 people were killed in the Iranian Rudbar earthquake. The emergency loss of the Japanese Hanshin earthquake in 1995 was as high as $96 billion. In order to resist and mitigate earthquake disasters, it is necessary to carry out seismic analysis and seismic design of building engineering structures.

According to the cause of the earthquake, the earthquakes are mainly divided into the following types:

(1) Tectonic earthquakes

Tectonic earthquakes are also known as "fault earthquakes". One type of earthquake is caused by faults in the earth's crust (or lithosphere, a few mantle sites on the lithosphere below the earth's crust). The earth's crust (or lithosphere) deforms during tectonic movement. When the deformation exceeds the rock's bearing capacity, the rock breaks. The long-term accumulated energy in the tectonic movement is quickly released, causing the rock to vibrate and form an earthquake. The scope of the spread is large and destructive. More than 90% of the world's earthquakes and almost all destructive earthquakes are tectonic earthquakes.

(2) Induced earthquakes

Earthquakes caused by engineering activities such as artificial blasting, mining and construction of reservoirs. The scope of influence is small and the earthquake intensity is generally small.

(3) Volcanic earthquake

Due to the active volcanic eruption, the magma rushed out of the ground and caused an earthquake. It mainly occurs in areas with volcanoes, which are rare in China. There are not many volcanic earthquakes, accounting for only 7% of the total number of earthquakes.

1. 1. 3　Causes of earthquakes

The cause of the earthquake is generally thought to be that the outermost layer of the earth is composed of some huge plates.

The six plates are the Eurasian plate, the American plate, the African plate, the Pacific plate, the Australian plate and the Antarctic plate. The depth of the plate extending downward is about 70~100km. Due to the convection of the mantle material, the plates also move with each other, and the tectonic movement of the plate is the root cause of the earthquake. The gap between the six plates is a seismic zone. The seismic zone refers to the concentrated distribution of earthquakes. There are three seismic zones on the earth (Pacific Rim volcanic belt, Mediterranean seismic belt, ridge seismic belt).

The Pacific Rim volcanic belt is located on the continental margins and islands on the Pacific Ocean. From the Andes Mountains on the west coast of South America, south to the southern tip of South America, the Malvinas Islands (Falkland Islands) to South Georgia; north to Mexico, the North American West Coast, the Aleutian Islands, Kamchatka, and the Kuril Islands to the Japanese archipelago; then divided into two, one to the southeast through the Mariana Islands, Guam to Yap Island, the other to the southwest through the Ryukyu Islands, Taiwan, the Philippines to Sulawesi, and the Mediterranean-Indonesia. After the confluence of the seismic zone, through the Solomon Islands, the New Hebrides, Fiji Island to New Zealand. Its basic position is the same as that of the Pacific Rim volcanic belt, but its influence range is slightly wider than that of the volcanic zone, and the continuous banding is also more obvious. This seismic zone concentrates 80% of the world's earthquakes, including a large number of shallow earthquakes, 90% of medium-source earthquakes, almost all deep-seismic earthquakes, and most of the world's most devastating earthquakes.

The Mediterranean seismic belt, from the Atlantic Ocean to the Azores, eastward through the Mediterranean Sea, Turkey, Iran, Afghanistan, Pakistan, the northern India, the western and southwestern borders of China, and through Myanmar to Indonesia, connected to the Pacific Rim. It traverses Europe, Asia and Africa, with a total length of more than 20, 000 kilometers, basically the same position as the east-west volcanic belt, but the banding characteristics are more distinct. The belt concentrates 15% of the world's earthquakes. Mainly shallow earthquakes and medium-source earthquakes, lack of deep earthquakes.

The ridge seismic belts are distributed at the axis of the global oceanic ridge, and all are shallow earthquakes, and the magnitude is generally small. In addition, there are some seismic belts with relatively small distribution within the mainland. Such as the East African Rift earthquake. The earthquakes are frequent in China's neighboring Pacific Rim seismic belt and the Mediterranean-Indonesian seismic belt. Destructive earthquakes have occurred in history and in the near future. The seismic zone is basically at the junction of the plates, and seismic zones are often associated with certain seismic structures.

China is located between the two major seismic belts of the world, the Pacific Rim and the Eurasian seismic belt, and is squeezed by the Pacific, Indian and Philippinesea plates.

Because the fracture of the rock formation is not a plane development, but a fracture zone composed of a series of fractures, the rock formation along the entire fracture zone cannot reach equilibrium at the same time. Therefore, after another strong earthquake (the main shock), the deformation of the rock forms. There are also constant sporadic adjustments that form a series of aftershocks.

Foreshocks: A few earthquakes are proceeded by smaller from the source area, which can used to predict the main shock.

Aftershocks: large earthquakes are sometimes followed by incredible numbers of them, which can results the damaged building destructed.

1. 2 Related concepts of earthquake

Earthquakes often occur in areas where the ground stress is concentrated and the structure is relatively fragile, that is, at the breakpoint or turning point of the original fault, and the intersection of different faults (Figure 1. 2).

Focus: The part of the earth where the fault is displaced and causes vibration of the surrounding medium.

Epicenter: The ground position directly above the source.

Epicentral distance: The horizontal distance from the ground to the epicenter.

Focal depth: The vertical distance from the source to the epicenter.

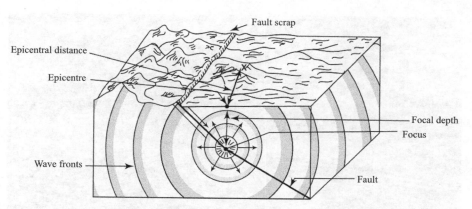

Figure 1.2 Earthquake

Fault: The crustal faults occur when the crust is subjected to stress, and significant relative displacements occur along the rock blocks on both sides of the fracture surface. The size of faults varies. The larger faults extend up to thousands of kilometers along the strike and cut through the crust downward. They are usually composed of many faults, called fault zones; the smaller ones are centimeters long and can be seen in rock samples.

Faults are classified as: (1) normal fault; (2) reverse fault; (3) strike-slip fault.

A normal fault is one in which the rock above the inclined fault surface moves downward relative to the underlying crust (Figure 1.3).

Figure 1.3 A normal fault

A reverse fault is one in which the crust above the inclined fault surface moves upward relative to the block below the fault (Figure 1.4).

A strike-slip fault, sometimes called a transcurrent fault or lateral fault, involves displacements of rock laterally, parallel to the strike (Figure 1.5).

Figure 1. 4   A reverse fault

Figure 1. 5   A strike-slip fault

Because faults play an important role in earthquake, it is found that the elastic re-bound theory is based on the in-depth study of faults. The elastic bounce theory was origi-nally proposed by M. F. Reid in 1906, which attributed the occurrence of tectonic earth-quakes to the gradual accumulation of strain in a particular region and the gradual increase in the elastic force that was subsequently stored. As previously discussed, the newly formed oceanic plate propels the continental plate, causing the mainland to drift. When the plates collide, they may be locked in place, that is, they may be prevented from moving due to frictional resistance along the boundaries of the plates. This results in an increase in stress along the edge of the plate until a sudden slip due to elastic bounce or rock fracture, resulting in a sudden release of strain energy, which may cause the earth's crust to break in one direction and form a fault. This is the origin of the earthquake. The gradual accumulation and subsequent release of stress and strain are described as elastic rebound. The elastic rebound theory assumes that the source of the earthquake is a sudden displacement of the ground on both sides of the fault, which is the result of the fracture of the crustal rock. The upper part of the crust and lithosphere is very hard and brittle. When the rock is deformed, it is actually slightly curved. However, it can withstand very light stresses with only slight bending or strain. The elastic rebound theory requires that the strain rapidly increase to the elastic limit of the rock. After this point, due to the formation of the fault, the crust ruptures and the curved rock quickly returns to its original shape, releasing energy stored in the form of bounce and severe vibration (elastic waves). These vibrations vibrate the ground, the maximum vibration can be felt along the fault. After the earthquake, the strain accumulation process at this modified interface between the rocks began again (Cachi earthquake, 1819). Most earthquakes occur at the boundaries of tectonic plates, called inter-plate earthquakes (Assam earthquake, 1950). Others occur inside the plate, away from the plate boundary (Latour earthquake, 1993), known as intraplate earthquakes. In both types, slip occurs during elastic strain accumulation and fracture.

According to tothe epicenter distance, the earthquake can be divided into:

(1) Local earthquake: an earthquake with an epicenter distance of less than 100 kilometers. The local earthquake recorded by the seismic station is generally based on the direct wave propagating in the earth's crust.

(2) Near-earthquake: an earthquake with an epicenter distance greater than 100 kilometers and less than 1000 kilometers. The first arrival wave recorded by the seismic station to the near earthquake is generally a diffraction wave, a reflected wave and a surface wave through the upper interface of the mantle.

(3) Teleseism: an earthquake with an epicenter distance of more than 1, 000 kilometers. The teleseism waves recorded by the seismic station contain nuclear-surface reflection waves propagating below the mantle, ground-penetrating waves and surface waves of the surface layer of the earth's crust.

According to the focus depth, earthquakes can be classified into shallow source earthquakes, medium source earthquakes and deep source earthquakes according to the source depth.

(4) Shallow earthquakes: earthquakes with a focal depth of less than 70 kilometers. It has the most frequent vibrations and the greatest impact on humans. If compared with the amount of energy released, 85% is a shallow earthquake. On the mainland, shallow earthquakes account for more than 95% of the earthquakes, so earthquake disasters are mainly caused by shallow earthquakes.

(5) The intermediate earthquakes: an earthquake with a focal depth between 70 and

300 kilometers, is called the intermediate earthquake. The intermediate earthquake generally does not cause disasters.

(6) Deep-seismic earthquakes: earthquakes with a focal depth of more than 300 kilometers, are called deep-seismic earthquakes. Most of them are distributed near the deep sea trenches in the Pacific Ocean, but at a distance. On June 29, 1934, an earthquake of magnitude 6. 9 occurred east of Sulawesi, Indonesia, with a focal depth of 720 km. It is the deepest known earthquake. Deep earthquakes generally do not cause disasters even if the magnitude is high.

1. 3   Seismic waves and ground motion

1. 3. 1   Seismic waves

Seismic energy propagates through the medium from the source to the surrounding in the form of waves, which is the seismic wave. A seismic wave is an elastic wave that includes body waves and surface waves.

Body waves: waves that propagate inside the earth are called body waves. Body waves have both longitudinal and transverse waves. The longitudinal wave is a compression wave (P wave) whose media particle motion direction is the same as the wave traveling direction (Figure 1. 6). The longitudinal wave period is short, the amplitude is small, the propagation velocity is the fastest; the transverse wave (S wave), the direction of the medium particle motion is perpendicular to the forward direction of the wave (Figure 1. 6). The transverse wave period is long, the amplitude is large, and the propagation velocity is second to the longitudinal wave, causing the ground to shake from side to side near the epicenter.

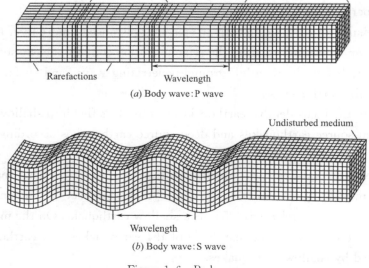

(a) Body wave: P wave

(b) Body wave: S wave

Figure 1. 6   Body wave

Surface wave: a wave propagating along the surface of the earth is called a surface wave. The surface wave has two forms: Love wave (L wave) and Rayleigh wave (R wave). When the Love wave propagates, the particle moves in a serpentine direction in the direction of the water perpendicular to the direction of advancement of the wave (Figure 1. 7). The surface wave velocity is the slowest, the period is long, the amplitude is large, and the attenuation is slower than the body wave. When the Rayleigh wave propagates, the particle moves in a reverse elliptical motion in the plane of the wave's forward direction and the surface normal (Figure 1. 7). And causing the ground to sway.

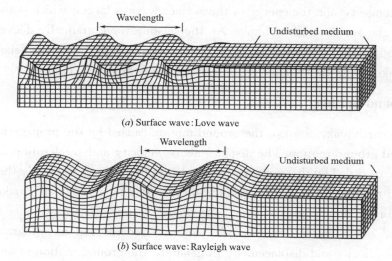

(*a*) Surface wave: Love wave

(*b*) Surface wave: Rayleigh wave

Figure 1. 7   Surface wave

The propagation velocities V_p and V_s of Pwaves and Swaves, respectively, are expressed as follows:

$$V_p = \sqrt{\frac{E(1-\nu)}{\rho(1+\gamma)(1-2\gamma)}} = \sqrt{\frac{\lambda+2G}{\gamma}} \tag{1-1}$$

$$V_s = \sqrt{\frac{E}{2\rho(1+\gamma)}} = \sqrt{\frac{G}{\rho}} \tag{1-2}$$

Where V_p——the velocity of the P wave;

  V_s——the velocity of the S wave;

  E——Young's modulus;

  γ——the Poisson ratio;

  ρ——the mass density of the medium;

  G——the shear modulus;

  λ——the Lame constant:

$$\lambda = \frac{\gamma E}{(1+\gamma)(1-2\gamma)} \tag{1-3}$$

Using Formula (1-1) and Formula (1-2), the ratio of V_p to V_s can be obtained as follows:

$$\frac{V_p}{V_s} = \sqrt{\frac{2(1-\gamma)}{1-2\gamma}} \qquad (1\text{-}4)$$

From Formula (1-1) and (1-2), $V_p = \sqrt{3} V_s$. Near the surface of the earth, $V_p = 5 \sim 7\mathrm{km/s}$ and $V_s = 3 \sim 4\mathrm{km/s}$. For general material, during earthquakes P waves travel faster than S waves.

In summary, the longitudinal wave first arrives at the time of the earthquake, the transverse wave is second, and the surface wave is the slowest; in terms of amplitude, the latter is the largest. When both the shear wave and the surface wave arrive, the vibration is the strongest, and the energy of the surface wave is large, which is the main cause of damage to the surface and buildings. As the seismic wave gradually decays during the process of propagation, with the increase of the epicentral distance, the seismic vibration gradually weakens and the damage of the earthquake is gradually reduced.

1.3.2 Ground motion

When an earthquake occurs, the ground motion caused by the propagation of seismic waves is called ground motion. The displacement, velocity and acceleration of the ground motion can be recorded by the instrument. People can understand and study the characteristics of ground motion based on the acceleration recorded by strong earthquakes. Using acceleration records, direct dynamic time history analysis and seismic response spectrum can be constructed for building structures. Acceleration records can be integrated to obtain ground motionvelocity and displacement. In general, the ground motion at one point has a component of six directions (three translational components and three rotational components) in space. Currently, a translation of the translational component is only obtained.

In fact, ground motion is the result of a combination of multiple seismic waves. Therefore, the recording signal of the ground motion is irregular. But through analysis, we can use several limited elements to reflect irregular seismic waves. For example, the maximum amplitude can be used to quantitatively reflect the intensity characteristics of ground motion; the spectrum analysis of seismic records can reveal the periodic distribution characteristics of ground motion; by defining and measuring the duration of strong earthquakes, the degree of ground motion can be investigated. The peak (maximum amplitude) of the ground motion, the spectrum and duration, are often referred to as the three elements of ground motion.

The peak value is the representation of the seismic vibration intensity of the ground motion amplitude, usually expressed as the peak value, such as peak acceleration and peak velocity. The peak value is the maximum value of the ground motion. The magnitude of the ground motion peak reflects the maximum intensity of ground motion at a certain moment in the earthquake process. It directly reflects the seismic force and the vibration energy generated by it and the magnitude of the earthquake deformation caused by the structure.

The ground motion spectrum characteristic refers to the influence of ground motion on the structural response characteristics of structures with different natural oscillation periods. A curve indicating the relationship between amplitude and frequency in a ground motion is collectively referred to as the spectrum. In earthquake engineering, Fourier spectrum, response spectrum and power spectrum are usually used to represent the definition of spectral characteristics, and the composition of various frequency components in ground motion is expressed.

The duration of strong ground motion is affected by earthquake damage and structure. The increase of duration improve the probability of occurrence of large permanent deformation. The longer the duration, the greater the reaction. This is the cumulative effect of earthquake damage. For the seismic design of general industrial and civil buildings, the magnitude (strength) of ground motion can be used. However, for major projects and special projects, only the amplitude is not good, and the duration needs to be considered. The seismic damage of the engineering structure is closely related to the three elements of ground motion.

1.4　Seismic metrics

1.4.1　Earthquake magnitude

The amount of energy released by an earthquake is measured by magnitude. The general methods for testing magnitudes are near-seismic magnitude, body-wave magnitude, surface-wave magnitude, moment magnitude, and Richter magnitude. Let's introduce the Richter magnitude and surface-wave magnitude.

Earthquake magnitude (Richter magnitude): The magnitude of the earthquake is measured in micrometers ($1\mu m = 10^{-6} m$) recorded by a standard seismograph (referring to a seismometer with a period of 0.8s, a damping coefficient of 0.8, and a magnification of 2800). Common logarithm of the maximum horizontal ground displacement A:

$$M = \log A \tag{1-5}$$

Where　　M——earthquake magnitude, proposed by Dr. Charles Richter of the United Kingdom in 1935.

　　　　A——maximum amplitude (μm) on the seismic time history curve.

$$M_L = \log A - \log A_0 \tag{1-6}$$

Where　　M_L——near-shock body wave magnitude (Richter magnitude).

　　　　A_0——that for a particular amplitude of earthquake selected as a standard.

In fact, earthquakes do not necessarily have seismometers at a distance of 100 km from the epicenter, and the above-mentioned standard seismographs are generally not used today. Therefore, for a seismic station with an epicenter distance of 100 km and a non-standard seismograph, the magnitude must be calculated according to the revised calculation formula.

According to the current instruments in China, the near-seismiccalculation (referring to the epicentral distance less than 1000km) is as follows:

$$M_L = \log A_\mu + R(\Delta) \tag{1-7}$$

Where A_μ———The horizontal maximum amplitude (μm) on the seismic curve.

$R(\Delta)$———the starting function of the variation from the Δ in the earthquake.

For the method of measuring the magnitude of theteleseismic (referred to as the epicentral distance greater than 1000km), the bulk wave method and the surface wave method are generally used. In China, the empirical formula for the surface wave magnitude is:

$$M_S = \log\left(\frac{A_u}{T}\right)_{max} + \sigma(\Delta) \tag{1-8}$$

Where M_S———near-shock surface wave magnitude;

A_u———The horizontal maximum amplitude (μm) on the seismic curve;

T———the period corresponding to A_u;

$\sigma(\Delta)$———the gauge function of the surface wave magnitude.

However, no matter what magnitude is used, the magnitudes determined by different earthquakes in different locations are often different due to the difference of seismic wave propagation medium, and the difference often reaches $0.3 \sim 0.5$, sometimes even more than 1.0.

$$\log E = 1.5M + 11.8 \tag{1-9}$$

【Example 1.1】 The formula for calculating Richter magnitude M is $M_L = \log A - \log A_0$, in which A is the maximum amplitude of seismic curve recorded by seismometer and A_0 is the corresponding amplitude of standard earthquake.

Questions:

Assuming that the maximum amplitude recorded by the seismometer is 1000, when the amplitude of standard earthquake is 0.001, what is the earthquake magnitude, and how many times the maximum amplitude of magnitude 9 earthquake is the one of magnitude 5 earthquake?

Solution:

The earthquake magnitude is $M_L = \log A - \log A_0 = \log 1000 - \log 0.001 = 6$

A M9 earthquake occurs, there will be $\log A_9 = M + \log A_0 = 9 + \log A_0$

A M5 earthquake occurs, there is $\log A_5 = M + \log A_0 = 5 + \log A_0$. Therefore, $A_9 = 10^{9+\log A_0}$, $A_5 = 10^{5+\log A_0}$

So, the answer to $\dfrac{A^9}{A^5} = 10^4 = 10000$

It can be seen from equations (1-1) and (1-5) that when the magnitude is increased by one level, the ground vibration amplitude is increased by 10 times, and the energy is increased by 32 times. The energy released by a seven-stage destructive earthquake is equivalent to nearly the energy of 30 atomic bombs of 20,000 tons.

The magnitude is usually divided into several levels (Table 1.1):

Class	Magnitude
Great	≥8. 0
Major	7. 0~7. 9
Strong	6. 0~6. 9
Moderate	5. 0~5. 9
Light	4. 0~4. 9
Minor	3. 0~3. 9

Magnitude Division　　　　　　　　　　　　　　　　**Table 1. 1**

Seismic intensity: seismic intensity is related to peak acceleration, velocity and duration. It refers to the average strength of a certain area's surface and various buildings affected by an earthquake. Seismic intensity is a measure of the extent of earthquake damage.

An earthquake has just a magnitude, but multiple earthquake intensity exists (Figure 1. 8).

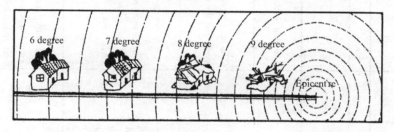

Figure 1. 8　　　Different intensity

The M8. 9 earthquake that occurred in Chile on May 22, 1960, is the largest earthquake recorded in the world.

1. 4. 2　Seismic intensity scales

The seismic intensity table has beenused for more than 400 years. Due to the lack of seismic observation instruments, the early intensity scales can only be formulated based on macroscopic phenomena of earthquakes, human perceptions, reflections of objects, influences and damages on the surface and buildings. Since the macro intensity scale does not provide quantitative data, it cannot be directly used for seismic design of engineering. With the advent of strong earthquake observation instruments, it is possible for people to use certain parameters of the recorded ground motion, such as peak acceleration and peak velocity to define the intensity, and thus a seismic intensity table that links the macroscopic intensity of the earthquake with the ground motion parameters. Table 1. 2 shows the China earthquake intensity issued by the National Seismological Bureau and implemented in 2008. It is worth pointing out that there is enough data to prove that the same intensity zone can have tens of times or even hundreds of times the acceleration value or velocity value corresponding to it, so the horizontal acceleration and velocity in Table 1. 2 cannot be used as the intensity. The indicators are evaluated and can only be used as a reference.

Seismic intensity scales

Table 1.2

Seismic intensity	Human feeling	House damage			Other earthquake damage	Horizontal ground motion	
		Type	Degree of damage	Average damage index		Peak acceleration (m/s²)	Peak velocity (m/s)
I	No sense	—	—	—	—	—	—
II	Some static persons have feeling in the room	—	—	—	—	—	—
III	Rare persons have feeling	—	Doors and windows squeak	—	Hanging motion	—	—
IV	Most people in the room, a few outsiders have feelings, and a few people wake up in their dreams	—	Doors and windows squeak	—	The hanging object swings obviously and the utensils rang	—	—
V	Most people in the room, most people in the outdoor have a feeling, most people wake up in their dreams	—	Doors and windows, roofs, roof trusses vibrate, lime soil falls, individual houses are plastered with fine cracks, and some tiles are dropped, and individual roof chimneys are dropped	—	The suspension is shaken a lot, and the unstable object shakes or falls over	0.31 (0.22~0.44)	0.33 (0.02~0.04)
VI	Most people are standing unsteadily, and a few people are hanging outdoors	A	A few moderate damage, most minor damage or basically intact		Furniture and articles move; cracks appear on river banks and soft soil, sandblasting occurs in saturated sand layers; light cracks in individual brick chimneys	0.63 (0.45~0.89)	0.06 (0.05~0.09)
		B	A few moderate damage, a few minor damages are mostly intact	0.00~0.11			
		C	A few moderate damage, most of them are basically intact	0.00~0.08			

Continued

Seismic intensity	Human feeling	House damage				Horizontal ground motion	
		Type	Degree of damage	Average damage index	Other earthquake damage	Peak acceleration (m/s²)	Peak velocity (m/s)
Ⅶ	Most people are stunned outdoors, cyclists have a feeling, and the driving occupants feel	A	Destroyed or severely damaged, most moderate or minor damage	0.09~0.31	Objects fall off the shelf; there is a landslide on the river bank; the sand is often sprayed with sand, and there are many ground cracks on the soft land; most of the independent brick chimneys are destroyed	1.25 (0.90~1.77)	0.13 (0.10~0.18)
		B	Few destruction, most serious or moderate damage				
		C	Individual destruction, few serious damage, most moderate or minor damage	0.07~0.22			
Ⅷ	Most people shake and bump, walking hard	A	Few destruction, most serious and moderate damage	0.29~0.51	Cracks appear on dry hard soil, most of the saturated sand layer is sandblasted; most independent brick chimneys are seriously damaged	2.5 (1.78~3.53)	0.25 (0.19~0.35)
		B	Individual destruction, few serious damage, most moderate or minor damage				
		C	A few severe or moderate damage, most minor damage	0.20~0.40			
Ⅸ	Man in action falls	A	Most severely damaged or destroyed	0.49~0.71	Cracks appear on dry hard soil, visible bedrock cracks, misalignment, landslides, landslides are common; independent brick chimneys are severely collapsed	5.00 (3.54~7.07)	0.25 (0.36~0.71)
		B	Few destruction, most serious and moderate damage				
		C	Minor damage or severe damage to moderate or minor damage	0.38~0.60			

Continued

Seismic intensity	Human feeling	House damage			Other earthquake damage	Horizontal ground motion	
		Type	Degree of damage	Average damage index		Peak acceleration (m/s²)	Peak velocity (m/s)
X	The cyclist falls, and the unstable person will fall off the ground and have a feeling of throwing away	A	Huge destruction	0.69~ 0.91	Landslides and earthquake breaks occur; arch bridges are destroyed; most independent brick chimneys are destroyed or destroyed from the roots	10.00 (7.08~14.14)	1.00 (0.72~1.14)
		B	Most destroyed				
		C	Most destroyed or severely damaged	0.58~ 0.80			
XI		A		0.89~ 1.00	Earthquake faults continue to be large, and a large number of landslides	—	—
		B	The vast majority of destruction				
		C		0.78~1.00			
XII		A			Dramatic changes in the ground, changes in mountains and rivers	—	—
		B	Almost completely destroyed	1.00			
		C					

Note: The "peak acceleration" and "peak velocity" given in the table are reference values, and the range of variation given in parentheses.

1. Quantifiers in the table. " Individual" is 10% or less; "minor" is 10% to 50%; "majority" is 60% to 90%; " most" is 80% or more.

2. When assessing seismic intensity, the I to V degrees should be based on the feelings of people on the ground and in the bottom house and other earthquake damage phenomena; VI~X degrees should be based on house damage, referring to other earthquake damage phenomena. when the damage degree of the house is different from the average earthquake damage index, the results of the earthquake damage assessment should be the main one, and the average earthquake damage index of different types of houses should be considered comprehensively; the XI degree and XII degree should comprehensively consider the house earthquake damage and the surface earthquake damage.

3. The results of seismic intensity assessment in the following three cases shall be adjusted appropriately:

(1) When assessing the seismic intensity using the response of the tall buildings and the response of the objects, appropriately reduce thelevel.

(2) When using the earthquake damage degree and the average earthquake damage index of buildings below or above VII to assess the seismic intensity, appropriately reduce or increase the level.

(3) When using the earthquake damage degree and the average earthquake damage index of buildings with particularly poor or particularly good buildings to assess the seismic intensity, appropriately reduce or increase thelevel.

Seismic zoning: according to geological structure data, historical earthquake law, strong earthquake observation data, and seismic hazard analysis method, the probability of overshoot of a certain intensity (or ground motion acceleration value) in each region within a certain time limitation can be calculated. Thus, the land can be divided into areas covered by different basic intensities, which is called seismic zoning.

1.5　Introduction to the earthquakes in China and World

1.5.1　Introduction to the Chinese historical earthquake

China had an understanding of the earthquake as early as the Warring States Period. In the year of A.D.132 (the Eastern Han Dynasty), Zhang Heng invented the Houfeng Earth Instrument, which is the world's first instrument for observing earthquakes and had a milestone in the history of world seismology. By the Ming and Qing Dynasty, the records of earthquake were more comprehensive, detailed and rich, and there were also explorations of earthquake genesis and earthquake prediction.

Since the 20th century, there have been nearly 800 earthquakes of magnitude 6 or a-bove in China, covering all provinces and municipalities except Guizhou and Zhejiang provinces and Hong Kong Special Administrative Region.

China's earthquake activity is high in (occur) frequency, strong in intensity, shallow in source and wide in distribution. It is a country with severe earthquake disasters. Since 1900, the number of people who died in the earthquake in China has reached 550, 000, accounting for 53% of the global earthquake deaths. Since 1949, more than 100 devasta-ting earthquakes have happened 22 provinces (autonomous regions and municipalities), including the eastern region. In 14 provinces, more than 270, 000 people were killed, ac-counting for 54% of all types of disasters in the country. The earthquake-affected area reached more than 300, 000 square kilometers, and the number of houses collapsed to 7 million. The severity of earthquakes and other natural disasters constitutes one of China's basic national conditions.

Statistics show that China's land area accounts for a fifteenth of the world's land area; China's population accounts for about one-fifth of the global population, or about 20 per-cent, not even twenty percent, however, China's land earthquakes accounted for one-third of the global land earthquakes, or about thirty-three percent, and the number of people who caused earthquake deaths has reached more than 1/2 of the world. Of course, this also has special reasons. First, China has a lot of people. Second, China's economy is developing, houses are not strong, and it is easy to collapse and easy to be broken. Third, China's earthquake activity is intense and frequent.

According to statistics, since the 20th century, the number of people killed by earth-quakes in China has accounted for 54% of people died in natural disasters including floods, mountain fires, mudslides and landslides in the country. From the point of view of the death of personnel, the earthquake is the head of the group; the biggest cause of economic losses is meteorological disasters (floods), and the economic losses caused by meteorological disasters are much larger than those of earthquakes.

The earthquakes that occurred in Chinaare described as following:

The Xingtai earthquake occurred in March 1966. On March 8th, a magnitude 6.8 earthquake occurred in Malan Village, Longyan County; on March 22, another 7.2 magnitude earthquake occurred in Dongwang Town, Ningjin County; on March 26, in Baijikou, Ningjin County, A magnitude 6.2 earthquake occurred again. The epicenters of these three major earthquakes are not far apart and migrate to the northeast. The earthquake activity opened the prelude to the 20th century earthquake climax in North China. After the Xingtai earthquake, China's earthquake science research entered the stage of earthquake prediction research.

The Haicheng earthquake occurred on February 4, 1975. The magnitude is 7.3, and the intensity of the epicenter is IX. The long axis of the extreme earthquake zone is northwest. This is China's first successful earthquake of magnitude 7 or above, and precautionary measures were taken before the earthquake, which greatly reduced the number of casualties. The foreshock activity of this earthquake is quite typical and is one of the important reasons for the success of this earthquake prediction. The medium-term forecast is based on "earthquake migration" "earthquake empty space" and leveling. The basis for short-term forecasts is a variety of macro precursors.

The Lancang and Gengma earthquakes occurred at 21:3 and 21:16 on November 6, 1988. The magnitudes of the earthquakes were 7.6 (Lancang) and 7.2 (Gengma), respectively, 120 km apart. The earthquake was only 13 minutes apart. The two counties were razed to the ground, injured 4,105 people, 743 people died, and the economic loss was 2.511 billion yuan.

The Kunlun Mountains earthquake occurred at 17:26 on November 14, 2001. It was the second largest earthquake in Chinese mainland since the founding of the People's Republic of China (8.1 magnitude), second only to the 8.6 magnitude earthquake in Tibet on August 15, 1950. Some parts of Qinghai, Sichuan and Gansu have a sense of earthquake. At 11 o'clock on the morning of November 15, there were several aftershocks in the earthquake zone, causing some houses in the main earthquake area of Qinghai to collapse. The Qinghai-Tibet Highway (National Highway 109) was broken, and a large crack zone appeared in Kunlun Mountain. The location of the earthquake was relatively remote and no casualties occurred.

The Yushu earthquake occurred at 7:49 on April 14, 2010, and the magnitude of the earthquake was 7.1. The epicenter of the earthquake was located near the county seat (Figure 1.9). 2,698 people were killed in the Yushu earthquake, causing economic losses of 800 billion yuan.

The Lushan earthquake in Yaan City, occurred at 8:02 on April 20, 2013, and the

Figure 1.9　Yushu earthquake in Qinghai

magnitude of the earthquake was 7.0. The focal depth is 13 kilometers. The epicenter is a-bout 100 kilometers from Chengdu. Chengdu, Chongqing and Shanxi have strong earth-quakes in Baoji, Hanzhong and Ankang. More than 99% of the houses in Longmen Town-ship, Lushan County, collapsed, and the hospitals and inpatient departments stopped working and stopped water and electricity. The affected population was 1.52 million yuan and the affected area was 12,500 square kilometers.

The Jiuzhaigou earthquake in Sichuan occurred at 21:19:46 on August 8, 2017, and the magnitude of the earthquake was 7.0 (Figure 1.10). The epicenter was located in Biman Village, 5 km west of the core scenic spot of Jiuzhaigou. At 20:00 on August 13, 2017, 25 people were killed in the earthquake, 525 people were injured, 6 people were lost, 176,492 people (including tourists) were affected, and 73,761 houses were dam-aged to varying degrees (including 76 collapsed).

Figure 1.10　Sichuan Jiuzhaigou earthquake

The focus is on the Tangshan earthquake and the Wenchuan earthquake, because the earthquakes in these two earthquakes were extremely serious, causing massive deaths and injuries and economic losses.

The Tangshan earthquake occurred at 3:42 on July 28, 1976 (Figure 1.11). The magnitude of the earthquake was 7.8, and the intensity of the epicenter is XI. At 18:43 on the same day, a magnitude 7.1 earthquake occurred in Min County, 40 kilometers from Tangshan, with an epicenter intensity of IX degrees. The Tangshan earthquake occurred in a densely populated and economically developed industrial city, causing extremely heavy losses. More than 240,000 people were killed and more than 160,000 people were injured in the earthquake. Tianjin, a large city adjacent to the Tangshan area, was also damaged by VIII to IX degrees. It has a wide range of influences, affecting 14 provinces, municipalities and autonomous regions. The radius of the damage range is about 250 kilometers. This earthquake occurred in areas with many observation networks, but short-term and imminent earthquake predictions failed. The main reason is that there is no fore-shock activity before the earthquake; other precursor phenomena appear later; the earthquake zone is an industrial city, which has a great interference to precursor observation. The study of source physics shows that the source displacement process of this earthquake is more complicated. Invest-igations have shown that pre-slip occurs before the earthquake.

Figure 1.11   Tangshan earthquake

The Wenchuan earthquake occurred at 14:28:04 on May 12, 2008, and the magnitude of the earthquake was 8.0 (Figure 1.12). Its intensity is large, the affected area is wide, and it is destructive. The aftershocks lasted for a long time. As of 7:00 on June 3, there were 10037 aftershocks in the Wenchuan earthquake zone. The earthquake hit a land of about 500,000 square kilometers in China. As of 10:00 on April 25, 2009, 69,227 people were killed, 374,643 people were injured, and 17,923 people were missing. Among them, 68,712 compatriots were killed and 17,721 compatriots were missing. A total of 5,335 students were killed or missing. Direct economic losses amounted to 845.1 billion yuan. This is the most influential earthquake since the founding of the People's Republic of China.

1.5.2   Introduction to the World earthquake

In addition to many earthquakes in China, foreign countries have also been endangered by earthquakes, and such damage has caused enormous damage to foreign economies and personnel. These damages have historical significance in understanding human earthquakes and earthquakes. The following is a brief introduction to the earthquake hazards in

Figure 1.12 Wenchuan earthquake.

other countries.

Japan's Kobe earthquake occurred at 5:46 am on January 17, 1995, and the magnitude of the earthquake was 7.2, killed more than 5,400 people, injured more than 34,000 people, and collapsed and damaged more than 190,000 houses. Billions of dollars, more than 500 fires occurred after the earthquake (Figure 1.13).

Figure 1.13 the Kobe earthquake in Japan and the earthquake in East Japan

The Indonesian earthquake occurred on December 26, 2004, which magnitude was 8.7 on the surface of the northwestern part of Sumatra, Indonesia (the magnitude of the moment in the United States was 9.3), and the epicenter was at 3.9°N and 95.9°E. This is the largest earthquake in the world in the past 40 years. The earthquake also caused a strong tsunami, causing huge casualties and property losses in the Indian Ocean islands and many countries along the coast. After the earthquake, there were many aftershocks, and the maximum aftershock reached 7 magnitude or above. The earthquake caused a huge tsunami. The tsunami caused a total of about 370,000 people deaths. The number of injured people is countless, and the direct economic losses are estimated to be billions of dollars.

The Chilean earthquake occurred at 14:34 on February 27, 2010, which was 8.8 magnitude earthquake struck Chile, 89km northeast of Concepcion, Chile. Located 339 kilometers southwest of Santiago, Chile, the source is 55 kilometers underground. The high-rise buildings in Santxiago, Chile, were shaken and power was interrupted in some areas. After that, several strong aftershocks occurred successively, which triggered a tsunami and affected many neighboring countries such as Argentina. The strongest aftershock was 6.9 magnitude on the Richter scale. The earthquake caused thousands of deaths and more than two million people were injured.

The Great East Japan Earthquake (Japan, 13: 46, March 11, 2011) is M9. 0. Before this earthquake, a lot of earthquakes occurred in the region. Among them, there were many earthquakes of magnitude 7. 2 on March 9. According to the Pacific Tsunami Warning Center, at 00: 45 on March 9, 2011 (Beijing time), a magnitude 7. 2 earthquake occurred in the eastern seas of Honshu Island (38. 3°N, 143. 3°E). The local earthquake in Tokyo, Japan, was level 3, and the earthquake lasted for about 1 minute. The earthquake has triggered a regional tsunami near the epicenter of the earthquake. At 11: 11 on the same day, the OFUNATO station on Honshu Island, Japan, detected a tsunami wave of 0. 54 meters. After the magnitude 9. 0 earthquake on March 11, the researchers suddenly discovered that the 7. 2 magnitude earthquake on March 9 was the foreshock of the 9. 0 magnitude earthquake, which broke the 7. 0+earthquake in the same area since the observation history. There may be a record of higher earthquakes.

The Mexico earthquake of M8. 2 occurred on September 13, 2017, 98 people were killed. A total of 2. 3 million people in Oaxaca and Chiapas were affected to varying degrees. Although the complex terrain of the two states made the rescue operation more difficult, the government's rescue coverage almost covered the entire disaster area. According to the bulletin issued by the National Earthquake Center of Mexico, the epicenter of the earthquake was located 137 kilometers southwest of Tonala, Chiapas, with a focal depth of 19 kilometers.

1. 6　Earthquake damage

Earthquake disasters are mainly manifested in three aspects: surface damage, building damage and various secondary disasters.

1. 6. 1　Surface damage

Earthquake caused by the earthquake has aground fissure, land trap, Sand liquefaction, landslide and so on.

(1) Ground fissure (Figure 1. 14a)

Under the action of strong earthquakes, ground fissure often occur. According to the different mechanism, the ground fissures are mainly divided into structural ground fissures and gravity ground fissures. The structural ground fissure is related to the geological structure. The length of thefault can reach several kilometers to several tens of kilometers, and the width of the slit is several meters to several tens of meters. Gravity ground fissures are formed by earthquakes due to the uneven geology and hardness and micro-geographic gravity, which are closely related to the stable state of the soil. Such cracks are widely distinguished in earthquakes. They are common in soft and humid soils such as roads, ancient rivers, banks, slopes, etc. They vary in shape and size, and their scale is smaller than structural fissure. The length of the seam can range from a few meters to several tens of meters. The meter is much deeper than 1~2m. Where the ground fissure passes can cause cracks in the house and damage to engineering facilities such as roads and bridges.

（2）Land trap （Figure 1.14b）

Under the action of strong earthquakes, the ground often suffers from earthquakes and damages buildings. Most of the subsidence occurs in the soft and compressive soil layer, such as large-area backfill soil, large cohesive soil and non-cohesive soil. The earthquake caused the friction between the soil particles to be greatly reduced or the chain structure to be destroyed, and the soil layer became dense, causing the ground to sink.

（3）Sand liquefaction （Figure 1.14c）

Sand liquefaction is also one of the main causes of ground deformation. When the vibration of the saturated sand caused by the earthquake reaches the overburden pressure, the loose saturated sand will completely lose the shear capacity. At this time, the groundwater will be sprayed from the ground, and a large amount of sediment will be entrained to form the so-called sandblasting phenomenon. If the buried sand layer is shallow, the bearing capacity of the foundation will drop sharply or even completely, which will cause the structure to sink and tilt rapidly, causing serious damage.

（4）Landslide （Figure 1.14d）

Under the action of strong earthquakes, river banks and steep slopes are often caused. In mountainous areas, rocks are often cracked and collapsed. Landslides can cause road closures, traffic disruptions, ruining houses and bridges, cloggingrivers, flooding villages, etc.

(a) Ground fissure　　　　　　　　　　(b) Land trap

(c) Sand liquefaction　　　　　　　　　(d) Landslide

Figure 1.14　Earthquake damage

1.6.2 **Destruction of buildings**

Destruction of buildings can be divided into the following categories:

(1) Structural loss of integrity and destruction

Under the action of strong earthquakes, due to the weak connection of components, node failure, failure of the support system, etc., the structure will lose its integrity and cause damage or collapse (Figure 1.15).

Figure 1.15   Structural loss of integrity and destruction

(2) Insufficient bearing capacity of load-bearing structure causes damage

During the earthquake, the ground motion caused the building to vibrate and generate inertial force, which not only increased the internal force of the structural components, but also changed the force properties, resulting in insufficient structural bearing capacity (Figure 1.16).

Figure 1.16   Insufficient bearing capacity of load-bearing structure causes damage

(3) Non-structural damage due to excessive deformation

Under the action of strong earthquakes, when the structure is subjected to excessive vibration deformation, sometimes the main structure does not reach the strength damage, but the non-structural members such as the retaining wall, partition wall, awning, and various decoration often fall off due to excessive deformation and cause damage or collapse (Figure 1.17).

Figure 1.17   Non-structural damage due to excessive deformation

(4) Damage caused by foundation failure

Under strong earthquakes, ground fissures, subsidence, landslides and liquefaction of foundation soils may cause foundation cracking, sliding or uneven settlement, causing foundation failure, loss of stability or loss of bearing capacity, and ultimately the overall tilt of the building (Figure 1.18). The fissures cause the collapse.

Figure 1.18   Damage caused by foundation failure

1.6.3   Secondary disasters

After the occurrence of a strong earthquake, the natural state and the original state of the society were destroyed, resulting in a series of earthquakes caused by landslides, mudslides, tsunamis, floods, plagues, fires, explosions, gas leaks, and the spread of radioactive materials (Figure 1.19). Disasters are collectively referred to as earthquake secondary disasters. Secondary disasters of earthquake are generally classified into fires, poisonous gas pollution, and bacterial pollution according to their causes. Among them, fire is the most common and serious in secondary disasters.

(a) Tsunami   (b) Mudslide

(c) Fire   (d) Explosion

Figure 1.19　Secondary disasters

1.7　Application of seismic technology in Chinese and foreign buildings

1.7.1　Application of seismic technology in Chinese architecture

There are many ancient buildings in China, and hundreds of years ago, some great architects thought of the methods of earthquake resistance and applied them precisely to the building. There are two famous buildings.

1. Zhaozhou Bridge

Zhaozhou Bridge is a very famous structure in China. Zhaozhou Bridge, also known as Anji Bridge, is located on the river in Zhao County, Hebei Province, across a river that is more than 37 meters wide (Figure 1.20). The bridge is built entirely of stone and called "Dashi Bridge", built in the Sui Dynasty from 595 to 605 A.D., it was designed and built by the famous craftsman Li Chun. It has a history of more than 1400 years and is the earliest and most intact ancient single-hole open-shoulder arch bridge in the world. Zhaozhou Bridge is the crystallization of the wisdom of the ancient working people and has created a new situation in the construction of Chinese bridges.

In 2015, it was awarded one of the top ten business cards in Shijiazhuang. It is China's first stone arch bridge. In the long years, despite numerous floods, wind and rain, ice and snow erosion and the test of eight earthquakes, it is safe and sound, standing on the river.

（1）Foundation site selection

The stratum here is made up of river water alluvial, the surface of the stratum is a

Figure 1. 20   Panoramic view of Zhaozhou Bridge

thick sand layer washed by water, and the following are fine stones, coarse stones, fine sand and clay layers. According to modern estimates, the formation here can withstand a pressure of 4. 5 to 6. 6kg per square centimeter, while the pressure on the ground of Zhaozhou Bridge is $5\sim6$kg per square centimeter, which can meet the requirements of the bridge. After selecting the bridge site, the foundation and abutment are built on top. After 1400 years of construction, the bridge base only sank by 5 centimeters, indicating that the stratum is very suitable for bridge construction.

(2) Construction process

Zhaozhou Bridge has the unique method of construction and the convenient construction repair. Li Chun (designer) took the materials locally and selected the hard blue-gray sandstone produced by the nearby counties as the bridge stone. In the method of stone arching, the longitudinal (shun-bridge) construction method is adopted, and the entire bridge is composed of 28 independent arch vouchers arranged side by side in the width direction; the arch thickness is 1. 03 meters, each of which is independent, separate operation, quite flexible. After each piece is completely closed, it will be built into an independent piece to complete the bridge, move the weight of the "eagle frame", and build another adjacent arch. This kind of masonry has many advantages. It can save the wood used for making the "eagle frame" and is easy to move. At the same time, it is good for the maintenance of bridge, that is, the stone of an arch is damaged, as long as the new stone is embedded, it can be partially repaired, it is not necessary to adjust the entire bridge.

(3) Building structure

Many strict measures have been taken to maintain the stability of the bridge, in order to strengthen the horizontal connection between the arches, the 28 arches form an organic whole, and the connection is tight and firm. Li Chun has taken a series of technical measures:

1) Five iron tie rods are evenly arranged along the bridge width direction on the main place, passing through 28 arch, and each tie rod has a semi-circular rod head exposed out-

side the stone to clamp 28 arch coupons and enhance its lateral connection. Each of the four small arches has an iron tie rod to play the same role (Figure 1. 21).

Figure 1. 21   Close view of Zhaozhou Bridge

2) On the outer side of several arches and on the small arches at both ends, a layer of stone is covered to protect the arch; on the two sides of the arch stone, there are 6 hook stones, and the main arch stone is hooked to make it firmly connected.

3) In order to make the adjacent arches tightly fit together, a " waist iron" is connected between adjacent arches on both sides, and adjacent stones between the roads are also arched. Wear a " waist iron" to chain the arches. Moreover, the sides of each arch are chiseled with fine to increase the friction and strengthen the lateral connection of each part. These measures have taken the entire bridge into a compact whole, enhancing the stability and reliability of the entire bridge.

The flat arch is the form of a flat curved arch, which not only increases the stability and load-bearing capacity of the bridge, but also facilitates the passage of humans and animals on the bridge and saves stones. Li Chun has designed two open shoulders on the shoulders of the big stone arch. The small arch enhances the flood discharging capacity of the bridge and reduces the weight of the bridge.

2. Yingxian Wooden Tower

Yingxian Wooden Pagoda, also known as Sakyamuni, is the full name of " Buddha Temple Sakyama". The Buddha Temple Sakyama is located in the northwestern Buddhist temple of Yingxian, Cangzhou City, Shanxi Province. Built in the Liao Dynasty (A. D. 1056), it was completed in 1195. It has been completed and is the highest and oldest wooden tower building in China. National key cultural relics protection unit, national AAAA level scenic spot. It is also known as the " Three Great Towers of the World" with the Leaning Tower of Pisa in Italy and the Eiffel Tower in Paris. In 2016, Sakyamta Tower was recognized by the Guinness Book of World Records as the tallest wooden tower in the world.

The Sakyama Tower is located between the mountain gate and the main hall on the north-south axis of the temple, and belongs to the layout of the "front tower apse". The tower is built on a four-meter-high platform with a height of 67. 31 meters and a bottom diameter of 30. 27 meters. It is a flat octagonal shape (Figure 1. 22).

Figure 1. 22   Yingxian Wood Tower

The first layer of the facade is heavy, the above layers are all single eaves, with a dark layer between each layer, which is actually nine layers. Because the ground floor is heavy and has a cloister, the tower looks like a six-storey eaves. Each layer is supported by two inner and outer wooden pillars. There are 24 pillars outside each layer, and there are eight pillars inside. There are many diagonal braces, beams, rafts and short columns used between the wooden pillars to form a complex beam wooden frame in different directions. The entire wooden tower shares 3000 cubic meters of red pine wood, weighing more than 2, 600 tons.

The bottom doors of the tower are opened respectively in the north and south. Flat railings are setted above the second floor and each floor is equipped with wooden stairs. There are four doors on each of the second to fifth floors, each with a wooden partition. Buddha statues are molded on all floors of the tower. The first floor is Sakyamuni, which is 11 meters high. There are six Buddha statues on the inner wall of the inner channel. King Kong, Heavenly King, and disciples are also painted on the two side walls of the door. The top of the tower is made of octagonal and pointed iron. There are wind chimes on each floor of the tower.

The design of the Sakyama tower boldly inherits the form of heavy buildings with national characteristics since the Han and Tang Dynasties. It makes full use of traditional architectural techniques and widely uses the arch structure. 54 kinds of arches are used on this tower. Each bucket arch has a certain combination form. The beams, squares and columns are integrated into one body, and each layer forms an octagonal hollow structure layer.

1. 7. 2 Application of seismic technology in foreign buildings

Japan is located in the seismic zone, so Japan's seismic technology has a long history of development and its theoretical foundation is great. Many countries' seismic design specifications are based on Japanese norms.

1. Rigid structure

Japanese architecture is good at using rigid structures to improve the seismic performance of buildings. It is understood that many high-rise apartments in Japan will be sold out shortly after the start of sales. An important factor is that most of these high-rise apartments have the same level of seismic design as high-rise office buildings. One of the tallest apartments in Japan, using the same steel pipes as the World Trade Center in New York, USA, ensures seismic strength. The steel pipe has a diameter of up to 800 mm and a thickness of 40 mm, and the steel pipe is also filled with high-strength concrete which is three times stronger than usual concrete.

2. Using rubber

Japanese architects generally use rubber to improve the seismic performance of buildings. For example, there is a shock-free apartment in Tokyo, Japan. Although it is as high as 93 meters, but the newly developed high-strength 16-ply rubber is used in its periphery, and the natural rubber system is used in the central part of the building. In this way, when the earthquake of which intensity is 6 occurs, the force of the building can be reduced to 1/2.

3. Sink Foundation

Japan has developed a " local buoyancy" seismic system that supports the entire building by the buoyancy of water on the basis of traditional seismic structures. According to Japanese media reports, this technology is to provide a water storage tank between the upper structure of the building and the foundation, so that the building is supported by the buoyancy of the water. The buoyancy of water bears about half of the weight of the building, which not only reduces the load-bearing load of the foundation, but also reduces the rigidity of the vibration-isolating rubber and reduces the rigidity of the supporting structure portion, thereby improving the insulation between the foundation and the foundation. When an earthquake occurs, the natural oscillation period is prolonged due to buoyancy, the time required for shaking once, the acceleration of the building sway is reduced. Therefore, a good seismic effect can be obtained in a soft zone such as the coastal area of the city. This technology not only has a good seismic effect, but also the water stored in the storage tank can be used for fire extinguishing in the event of a fire, or as a temporary domestic water after an earthquake. More importantly, the cost of this system is not high. Take the eight-story hospital as an example, the cost is about 2% higher than the ordinary seismic system.

4. Sliding body foundation

Improving the seismic performance of buildings with a "sliding body" foundation. This technology is suitable for single-family, old buildings, and can effectively carry out earthquake protection of ancient buildings. This technique is to add a spherical bearing or sliding body between the building and the foundation to form a rolling support structure to mitigate the shaking caused by the earthquake. Japan has already implemented such repairs for ancient buildings such as the National Museum of Western Art.

5. Spring foundation

In order to prevent earthquakes, the Japanese can rack their brains. The construction department of Kashima, Japan, discovered a construction method for a shock-proof building: the spring was installed on the foundation of the building. The shock-proof building is characterized by the installation of a spring between the foundation of the building's foundation and the main part of the building, leaving the building in a floating state. Since the spring is an intermediary that absorbs earthquakes and other vibrations, the building itself is not subject to too strong impact, no matter how the foundation shakes. Experiments have shown that after the 6~7 magnitude earthquake is cancelled by the spring, the vibration will be reduced to 1/10.

6. Room wrapped in "bandage"

In Japan, where earthquakes occur frequently, a new type of cheap anti-seismic reinforcement technology has emerged. This technology uses resin materials as a seismic "bandage" to wrap the building pillars, thus preventing the pillars from collapsing during an earthquake.

Japan's "Asahi Shimbun" reported that this kind of anti-seismic reinforcement technology developed by the Japanese "Construction Quality Assurance Research Institute" researcher is called "SRF process. "The anti-vibration "bandage" is made of resin fiber and is shaped like a seat belt. During the construction, the anti-vibration "bandage" is coated with adhesive and wrapped on the pillars of the building. When an earthquake occurs, the pillars will not collapse even if there is internal damage, which ensures the living space of the people inside the building.

Specifically, taking a four-story building with 12 classrooms on each floor as an example, the reinforcement project usually cost 50 million yen (1 US dollar equivalent to 105 yen) to 100 million yen in Japan, using new technology. After that, it only costs about 5 million yen. If it is a wooden building, it only costs hundreds of thousands of yen. The construction is also quite simple. This new technology has been used in more than 250 construction projects, including the railway, hospitals and about 40 school buildings.

1. 8   Chinese building seismic precautionary target

1. 8. 1   Seismic precautionary intensity

Basic intensity: the basic intensity of an area is the maximum seismic intensity that may be experienced under normal site conditions for a certain period of time in the future. A certain period in the future refers to the period of use of buildings that have no special regulations or requirements (such as 50 years, 100 years...) from the time when the basic intensity is promulgated. General site conditions refer to standard foundation soil, general topography, landform, structure, hydrogeology and other conditions. Therefore, the basic intensity is the most common earthquake intensity in the region that may be encountered in a certain period of time in the future. Studies have shown that within 50 years, under normal site conditions, it may be subject to a 10% intensity of the probability of surpassing.

Seismic precautionary intensity: the basic intensity is adopted under normal circumstances. However, it must be determined according to the size of the city where the building is located, the type and height of the building, and the local seismic precautionary community plan. The seismic intensity that is approved as a seismic precautionary in a region according to the authority of the state is called seismic precautionary intensity.

1. 8. 2 Seismic precautionary classification and seismic precautionary standard

Due to the different functional characteristics of buildings, the social and economic consequences of earthquake damage are different. For buildings with different uses, different seismic precautionary standards should be used. According to the importance of buildings and the severity of the damage caused by earthquakes, the seismic design codes for buildings in China areclassified into four categories.

Building precautionary classification:

(1) Category A buildings (special precautionary): refers to major construction projects and buildings where an important secondary disaster may occur during an earthquake. The damage caused by such buildings is serious. Generally, the inpatient department, medical technology building, and outpatient department of large hospitals should be classified into Category A.

(2) Category B buildings (key precautionary): means that the function of the building cannot be interrupted by earthquake or the building needs to be restored as soon as possible. Such as urban lifeline projects, generally including water supply, power supply, transportation and so on.

(3) Category C buildings (standard precautionary): refers to general industrial and civil buildings in addition to buildings A, B and D.

(4) Category D buildings (moderate precautionary): refers to secondary buildings, including general warehouses, auxiliary buildings, etc.

Building precautionary standards (Table 1. 3):

(1) Category A buildings in the 6~8 degree zone are calculated according to the increase in the intensity of the precautionary. The seismic action and seismic measures should be determined according to the criteria for increasing the intensity of the zone's precautionary. When it is 9 degree, it should meet the higher requirements than the 9 degree seismic precautionary.

(2) Category B buildings are seismically calculated according to the intensity of the

precautionary, but they have been considered for the seismic construction measures.

（3）Category C buildings are subjected to seismic calculations and seismic structures in accordance with the intensity of the precautionary.

（4）Category D buildings shall be subjected to seismic calculation according to the intensity of the precautionary, but the seismic construction measures may be appropriately reduced（not falling at 6 degree）, and when the seismic precautionary intensity is 6 degree, unless otherwise specified, for B and C, D buildings can be calculated without seismic action.

Building precautionary standards　　　　　　　　　　　　　**Table 1. 3**

Category	Seismic action	Seismic measures	For instance
A	↑	↑	Inpatient department
B	—	↑	Hospital
C	—	—	Civil buildings
D	—	↓	Warehouses

1. 8. 3　Seismic precautionary target

Seismic intensity in an area is considered as a random variable, as shown in Figure 1. 23.

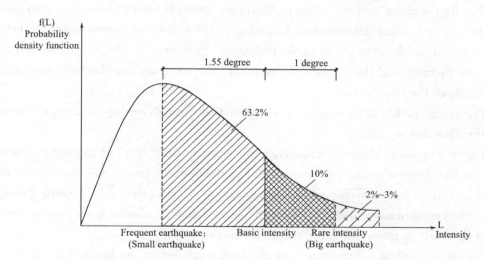

Figure 1. 23　Earthquake probability density function

Frequent earthquake intensity refers to the intensity, the probability of exceeding which is 63% under normal site conditions in the next 50 years period. The relative earthquake is called frequent earthquake. It is equivalent to an earthquake in 50 years.

Basic earthquake intensity refers to the intensity, the probability of exceeding 10% that may be encountered under normal site conditions in the next 50 years period. It is equivalent to an earthquake in 475 years.

Rare earthquake intensity refers to intensity, the probability of surpassing 2% to 3% under normal site conditions in the next 50 years period. It is equivalent to an earthquake that occurred once in 1600~2500 years.

Corresponding to the above design criteria, the Chinese code, *Code for Seismic Design of Buildings* GB 50011—2010, clearly puts forward three levels of seismic precautionary target:

First level: when subjected to earthquakes that are less than the intensity of the area's precautionary, buildings are generally not damaged or repaired and can continue to be used.

Second level: buildings may be damaged when subjected to an earthquake equivalent to the intensity of the area's precautionary, but they can continue to be used after general repairs.

Third level: when subjected to the rare earthquake that is higher than the intensity of the precautionary in the region, the building does not suffer serious damage to life-threatening.

1. 8. 4 Two-stage design method

In order to achieve the above-mentioned three-level seismic precautionary target, the Chinese code for buildings adopts the two-stage design method:

The first stage of design: when encountering the first level of intensity, the structure is in the stage of elastic deformation. According to the load effects combined with the one of seismic action which corresponds to the precautionary intensity, the bearing capacity of the structural member and the elastic deformation of the structure are checked to meet the requirements of the first level.

The requirements of the second level is satisfied through conceptual design and seismic structural measures.

The second stage design: when encountering the third level of intensity, the structure is in the stage of inelastic deformation. The elastoplastic inter-layer displacement check should be carried out according to the seismic action effect of the rare intensity corresponding to the precautionary intensity, and the corresponding seismic construction measures should be taken to meet the third level requirement.

For most building structures, only the first stage design can be carried out. For some buildings with special requirements or irregular structures, in addition to the first stage design, the second stage design should be carried out.

Summary:

Mankind has been bearing the brunt of the furies of nature since time immemorial, and has never given up the quest to develop tools and technologies to meet the challenges. One cannot imagine a greater destructive agent than an earthquake. This chapter is dedicated to first, understanding the cause, nature, effect, and consequences of an earthquake; second, presenting available methodologies to quantify this gigantic force; and third, providing Chinese ancient architecture and Chinese building seismic precautionary target.

Exercises

1. 1   What is the seismic wave? What kinds of waves are there?

1. 2   What are the differences between the earthquake magnitude and the earthquake intensity?

1. 3   How to classify the seismic precautionary buildings? What is the precautionary standard?

1. 4   What are the basic intensity, the precautionary intensity, the minor earthquake intensity, the moderate earthquake intensity and the major earthquake intensity?

1. 5   What are the frequent earthquake, the precautionary earthquake and the rare earthquake?

1. 6   What is the seismic precautionary target?

1. 7   How to achieve the "three levels" of seismic precautionary?

Chapter 2   Seismic Concept Design of Building

The seismic concept design refers to the process of designing the overall layout of the building and structure and determining the detailed structure according to the basic design principles and design ideas formed by earthquake disasters and engineering experience. The purpose is combining the uncertainty and regularity of the earthquake and its effects. The designer should focus on the overall reflection of the structure flexibly apply the seismic design criteria according to the structural failure mechanism and the destruction process at the beginning. The essence of good structural design (such as grasping the overall structure, structural system, bearing capacity and stiffness distribution, structural ductility, etc.), taking into account the details of key parts, and strive to eliminate the weak links in the structure (or develop clear seismic performance targets for key locations), fundamentally guarantee the seismic performance of the structure.

Previous earthquakes have shown that the intensity specified in Chinese seismic zoning maps is highly uncertain, and the seismic design is still in the exploration stage. The earthquake theory needs to be improved. Paying attention to the design of seismic design of buildings is the key point that should be grasped in seismic design. In a sense, the seismic concept design is also a remedy for the imperfect seismic theory.

Structural concept design does not reject complex structural design, but requires explicitly when dealing with complex structures: what is the best choice for structural design? what are the consequences of using unreasonable structural solutions or structural arrangements? What remedies or strengthening measures are needed? The objective evaluation of the rationality and effectiveness of these measures is carried out to ensure the realization of structural performance objectives and to ensure the safety of the houses. The conceptual design of the structure does not refer to the empty teaching by hand, but the tangible work with rich connotations.

2.1   Site and foundation

2.1.1   Building site selection

When selecting the construction site of building, the favorable, general, unfavorable and dangerous seismic sections shall be comprehensively evaluated according to the engineering needs, seismic activities, engineering geology and seismic geological data. As for the unfavorable sections, requirements on avoiding shall be proposed; if not, effective measures shall be taken. Category A and B buildings must not be constructed and the Category C buildings shall not be constructed in the hazardous sections. The division of various locations is given in Table 2. 1.

Division of favorable, common, unfavorable and hazardous sections   Table 2. 1

Section type	Geological, topographical and geomorphic description
Favorable section	Stable rock bed, stiff soil, or wide-open, even, compacted and homogeneous medium-stiff soil
Common section	Section not belonging to the favorable, unfavorable and hazardous sections
Unfavorable section	Soft soil, liquefied soil, stripe-protruding spur, lonely tall hill, steep slops, steep step, river bank or boundary of side slops, soil layer having obviously heterogeneous cause of formation, rock character and state in plane (including abandoned river beds, loosened fracture zone of fault, and hidden swamp, creek, ditch and pit, as well as base formatted with excavated and filled), plastic loess with high moisture, ground surface with structural fissure, etc.
Hazardous section	Places where landslide, collapse, land subsidence, ground fissure and debris flow may occur during the earthquake, as well as the positions in causative fault where ground dislocation may occur.

The soil layer of the site is not only the support of the building, but also the medium for transmitting seismic waves. The soil conditions of the site soil will affect the magnitude and characteristics of the surface ground motion. The site soil has a filtering effect and amplification effect on the bedrock wave, so that the earthquake of the hard site. The short-term cycle is dominant, while the weak site is dominated by long periods and directly affects the degree of damage to buildings. During the seismic action, the foundation transmits the ground motion to the upper structure while also transmitting the seismic forces received by the building back to the foundation.

Reasonable selection of sites that are favorable to earthquakes can avoid unfavorable sections and constructions that are not in dangerous areas, avoiding surface displacement and ground fissure caused by earthquakes, uneven subsidence of foundation soil, landslides and saturated silt, and liquefaction of sand. Therefore, choosing the right site is a very effective, reliable and economical seismic measure in structural seismic design.

2. 1. 2   Adjustment of seismic countermeasures for different sites

As can be seen from Table 2. 2, the site class is determined by the equivalent shear wave velocity of soil layer and the cover layer thickness of the site.

Cover layer thickness (m) of various building sites   Table 2. 2

Shear wave velocity of rock or equivalent shear wave velocity of soil (m/s)	Site class				
	I_0	I_1	II	III	IV
$v_{se} > 800$	0				
$800 \geqslant v_{se} > 500$		0			
$500 \geqslant v_{se} > 250$			< 5	$\geqslant 5$	
$250 \geqslant v_{se} > 150$			< 3	$3 \sim 50$	> 50
$150 \geqslant v_{se} > 500$			< 3	$3 \sim 15$	$15 \sim 80$	> 80

Note: v_s refers to the shear wave velocity of rock.

If the building site is Class I, it is still allowed to adopt details of seismic design for the Category A and B buildings according to the requirements of local seismic precautionary intensity and adopting details of seismic design for the Category C buildings according to the requirements of one grade less than the local seismic precautionary intensity, however, as for the seismic precautionary intensity 6, buildings shall still be adopted with details of seismic design according to the local seismic precautionary intensity (Table 2.3).

Adjustment of precautionary standards

when determining the details of seismic design for Class I building sites　　　Table 2.3

Building category	Seismic precautionary intensity in this region			
	6	7	8	9
A and B	6	7	8	9
C	6	6	7	8
D	6	6	7	8

If the building site is Class III or IV, where the design basic acceleration of ground motion is 0.15g or 0.30g, unless otherwise stated in this code, buildings should be adopted with details of seismic design according to the requirements of buildings belonging to each precautionary category respectively for seismic precautionary intensity 8 (0.20g) and 9 (0.40g).

According to the *Code for Seismic Design of Buildings* GB 50011—2010 and *Standard for Classification of Seismic Protection of Building Construction* GB 50223—2008, when determining seismic measures and the details of seismic design, the adjustments to the precautionary standards can be summarized as follows (Table 2.4):

Adjustment of precautionary standards

when determining the details of seismic design for different building sites　　　Table 2.4

Seismic precautionary category for structures	Seismic precautionary intensity in the region		Determining the standards for seismic measures				
			Class I site		Class II site	Class III, IV site	
			Seismic measures	Details of seismic design	Seismic measures	Seismic measures	Details of seismic design
Category A buildings	6	0.05g	7	6	7	7	7
	7	0.10g	8	7	8	8	8
		0.15g	8	7	8	8	8+
	8	0.20g	9	8	9	9	9
		0.30g	9	8	9	9	9+
	9	0.40g	9+	9	9+	9+	9+

Continued

Seismic precautionary category for structures	Seismic precautionary intensity in the region		Determining the standards for seismic measures				
			Class I site		Class II site	Class III, IV site	
			Seismic measures	Details of seismic design	Seismic measures	Seismic measures	Details of seismic design
Category B buildings	6	0.05g	7	6	7	7	7
	7	0.10g	8	7	8	8	8
		0.15g	8	7	8	8	8+
	8	0.20g	9	8	8	9	9
		0.30g	9	8	9	9	9+
	9	0.40g	9+	9	9+	9+	9+
Category C buildings	6	0.05g	6	6	6	6	6
	7	0.10g	7	6	7	7	7
		0.15g	7	6	7	7	8
	8	0.20g	8	7	8	8	8
		0.30g	8	7	8	8	9
	9	0.40g	9	8	9	9	9
Category D buildings	6	0.05g	6	6	6	6	6
	7	0.10g	6	6	6	6	6
		0.15g	6	6	6	6	7
	8	0.20g	7	7	7	7	7
		0.30g	7	7	7	7	8
	9	0.40g	8	8	8	8	8

Note: 1. "8+" It can be understood as "should meet the higher requirements than the 8 degree seismic precautionary".

2. "9+" It can be understood as "should meet the higher requirements than the 9 degree seismic precautionary".

3. The 《Code for Seismic Design of Buildings》 needs to be implemented in accordance with the relevant special regulations.

4. Seismic measures: the seismic design contents except earthquake action calculation and resistance calculation, including the details of seismic design, such as internal force adjustment measures.

5. Details of seismic design: all the detailed requirements must be taken for the structural and seismic concept design principles, requiring no calculation generally.

The *Standard for Classification of Seismic Protection of Building Constructions* stipulates that key fortifications should be used to improve seismic measures without increasing seismic action. The norms of some countries only increase the seismic effect (10% to 30%) without increasing seismic measures. There are differences in the concept of fortification: improving seismic measures and focusing on the use of financial and material resources to increase the seismic capacity of weak structural parts is an economical and effective method. Only by improving the seismic action, all components of the structure are fully increased. The effect of increased investment is not as good as the former.

2. 1. 3　Foundation design requirements

Foundation of one same structural unit should not be built on the bases with entirely different features. When an earthquake occurs, the foundation acts both to transmit ground vibrations to the engineering structure and to transmit the seismic forces received by the building to the foundation. The base bottom surface is preferably at the same elevation, and the foundation of the same structural unit is susceptible to seismic damage due to differences in ground motion transmission on foundations of distinct nature.

One same structural unit should not be adopted with natural base and pile foundation partially, if different types of foundations are adopted or the buried depth of foundation is different obviously, corresponding measures shall be taken at the relevant positions of foundation and superstructure according to the differential settlement of these two parts of base foundations under earthquake.

It is not advisable to use a natural foundation and a pile foundation for the same structural unit. For the high-rise buildings without the seismic joints in the main building and the podium, because the long-term maximum settlement of the podium is generally small, when the main building adopts the pile foundation, the podium can be allowed to adopt the natural foundation, but it is necessary to carefully analyze the different foundations in the earthquake. However, it is necessary to carefully analyze the difference of deformation of different foundation under earthquake and the difference of seismic response of superstructure and take corresponding measures (for example, sparse pile can be added to the natural foundation part).

For the base consisted of soft clay, liquefied soil, newly filled soil or extremely non-uniform soil, the corresponding measures shall be taken according to the differential settlement of base under earthquake and other adverse impacts such as the add of the ring beams.

2. 2　Regularity of building shapes

2. 2. 1　Basic requirements for building shapes

Whether on the building plane or on the elevation, the keypoint is to make the distribution of mass, stiffness and ductility uniform, symmetrical and regular, and avoid the sudden changes. The structure is symmetrical, which is beneficial to alleviate the torsional effect of the structure. The structure with regular shape is easy to coordinate the vibration of each part during the earthquake, and the possibility of reducing the stress concentration is beneficial to earthquake resistance. Therefore, in the seismic design, the regular structure and the irregular structure should be strictly distinguished, and the corresponding reinforcement measures should be taken for the irregular structure.

2.2.2   Plane Assembly

1. Total requirements

Generally speaking, the vertical component of seismic force is small, only $1/3 \sim 2/3$ of the horizontal component. In many cases such as $6 \sim 8$ degree zone, the influence of horizontal seismic force can be mainly considered. Correspondingly, the general layout of structures is mainly the layout of lateral resistant structures resisting horizontal forces. The overall layout of the structure is a key issue affecting the seismic performance of the building. The plane layout of the structure must be favorable for resisting horizontal forces and vertical loads. The force is clear and the force transmission is direct. The distribution of the planar shape and the lateral force structure of each structural unit of the building should strive for simple regular uniform symmetry, and reduced torsion influences (Figure 2.1).

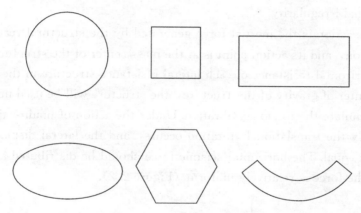

Figure 2.1   Simple regular plane

In the construction of the seismic zone, the favorable planar shape is circular, square, rectangular, and the regular polygon and the ellipse are also favorable planar shapes. However, In engineering, due to the various requirements such as construction land, urban planning, architectural art and functions, buildings can not necessarily be squares and circles. L-shape, T-shape, U-shape, H-shape will inevitably appear. For non-square, non-circular architectural planes, not necessarily irregular buildings, there is a question that how to identify flat-structured buildings.

2. Irregular plane

The *Code for Seismic Design of Buildings* GB 50011—2010 distinguishes between plane regularities and irregularities, and divides the plane irregularities into torsional irregularity, uneven irregularity, and partial discontinuity of floor slab, and some quantitative reference limits are shown in Table 2.5. These indicators are reference values for conceptual design rather than rigorous values and require a comprehensive judgment when used.

Main type of plane irregularity   Table 2. 5

Type of irregularity	Definition and reference index
Torsional irregularity	Under the action of specified horizontal force, the maximum elastic horizontal displacement or (storey drift)of storey is larger than 1. 2 times of the elastic horizontal displacement (or storey drift)at both ends of storey
Uneven irregularity	The sunken size of plane is larger than 30% of the overall size in the corresponding projection direction
Partial discontinuity on floor slab	The size of floor slab and the rigidity of plane change rapidly, for instance, the effective width of floor slab is less than 50% of the typical width of floor slab at this storey, or the opening area is larger than 30% of the floorage of this storey or great split-storey exists

Note: 1. " effective width of floor slab "refers to the actual seismic transmission width of the considered location, that is, the actual width of the slab after deducting the relevant opening.

     2. " typical width of floor slab " refers to the floor area as the floor width of most areas.

(1) Torsional irregularity

The seismic action is the inertial force generated by the structural reaction caused by the ground motion, and its action point is at the mass center of the structure. If the resultant points of horizontal resistance of each lateral resistance structure in the structure coincide with the center of gravity of thestructure, the structure will be used under the ground and will not stimulate the torsional vibration. Under the action of unidirectional horizontal earthquake, only the translational vibration occurs, and the lateral displacement of each layer member is equal. The horizontal seismic force should be distributed according to the stiffness, and the force is relatively uniform (Figure 2. 2).

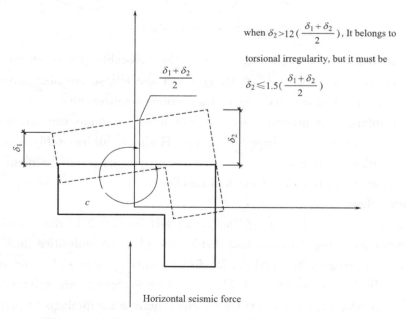

when $\delta_2 > 12\left(\dfrac{\delta_1+\delta_2}{2}\right)$, It belongs to torsional irregularity, but it must be

$$\delta_2 \leqslant 1.5\left(\dfrac{\delta_1+\delta_2}{2}\right)$$

Figure 2. 2   Example of torsional irregularity of building plane

The asymmetric structure does not coincide with the center of mass and the center of stiffness. Even under the action of a one-way horizontal earthquake, the torsional vibration is excited and the translational torsional coupled vibration is generated. Due to the effect of torsional vibration, the amount of lateral displacement of the member away from the center of stiffness is significantly increased. The horizontal seismic shear load is also significantly increased, and it is easy to cause serious damage due to exceeding the allowable resistance and deformation limit. It even caused the overall structure to collapse due to failure of one side member. In order to minimize the torsional effect, the distance between the center of the structural mass and the center of stiffness should be minimized.

For earthquake-resistant buildings, even if the structural arrangement is symmetrical, the mass distribution of the building is difficult to achieve uniform distribution, and the deviation of the center of mass and the center of the gravity of the structure is inevitable. Moreover, the ground motion is not only a translational motion, but also often accompanied by a rotational component. It is possible to have torsional vibrations in the structure during the earthquake. Therefore, in addition to the requirement of symmetry in the structure, it is also desirable to have a large torsional stiffness. The seismic wall with large lateral stiffness is preferably along the exterior wall of the building. The surrounding is arranged to increase the overall torsional stiffness of the structure. At the same time, special attention should be paid to the position of the reinforced concrete wall and the reinforced concrete core tube with great torsional stiffness, so as to be centered and symmetrical in the plane. In addition, the seismic wall should be arranged along the perimeter of the house, so that the structure has greater torsional stiffness and greater resistance to overturning. The lateral members of the same floor should have approximately the same stiffness, bearing capacity and ductility. The cross-sectional dimensions should not be too large, so as to ensure that the members can be jointly stressed, and avoid being broken by various forces in the earthquake. Such examples have occurred in previous earthquakes.

When the shape of the building and its components are irregularly arranged, the seismic action calculation and internal force adjustment shall be carried out.

For buildings with irregular planes and vertical regular, the spatial structure calculation model shall be adopted and the following requirements shall be met:

When the irregularities are reversed, the torsional effects shall be included. The maximum value of the horizontal displacement or the inter-layer displacement of the lateral force members at both ends of the floor should not exceed 1.5, when it under the action of the specified horizontal force with occasional eccentricity. When the maximum interlayer displacement is much smaller than the specification limit, it can be appropriately relaxed

(2) Uneven irregularity

When the shape of the building and its components are irregularly arranged (Figure 2.3), the seismic action calculation and internal force adjustment shall be carried out as follows, and effective seismic construction measures shall be taken for the weak parts.

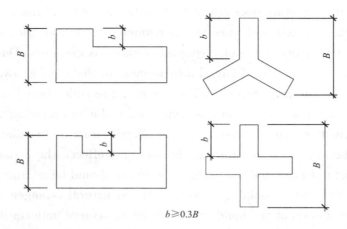

$b \geqslant 0.3B$

Figure 2.3 Example of uneven irregularity of building plane

For the buildings with irregular planes and vertical regular, the spatial structure calculation model should be adopted and should meet the following requirements:

1) In case of the uneven irregularity or the partial discontinuity of floor slab exists, the calculation model meeting the practical rigidity changes in the floor level shall be adopted. For high intensity or relatively big degree of irregularity, the influence of the local deformation of floor slab also should be taken into account.

2) For the buildings, with dissymmetrical plane and uneven irregularity or partial discontinuity, the torsional displacement may be calculated in blocks depending on actual situation, and the position with large torsion shall be adopted with local internal force enhancement coefficient.

【Example 2.1】 A six-story office building adopts the cast-in-place reinforced concrete frame structure. The elastic storey drift layer under seismic action is shown in Table 2.6.

The elastic storey drift layer under seismic action   Table 2.6

	X-direction storey displacement value		Y-direction storey displacement value	
	maximum(mm)	average(mm)	maximum(mm)	average(mm)
1	5.0	4.8	5.45	4.0
2	4.5	4.1	5.53	4.15
3	22.0	2.0	3.10	2.38
4	1.9	1.75	3.10	2.38
5	2.0	1.8	3.25	2.4
6	1.7	1.55	3.0	2.1

Question:

Judging the structural torsion regularity.

Solution:

X direction: $\dfrac{\text{Maximum displacement}}{\text{Average displacement}}$, both are less than 1.2.

Y direction: 1 to 6 floors $\dfrac{\text{Maximum displacement}}{\text{Average displacement}}$, respectively: 1.36, 1.33, 1.3, 1.3, 1.35, 1.43, both are greater than 1.20.

Since the ratio of the maximum displacement to the average displacement is greater than 1.2, it belongs to torsional irregularity structure.

【Example 2.2】 A building located in the 7 degree seismic precautionary area, it is a seven-storey complex building, 28m high. 1st to 3rd floors for shopping malls, 4th floor for conversion floor, 5th to 7th floors for hotels. In the middle of the shopping malls of the 1st to 3rd floors, there is a shared space with a large hole of 24m×10m. The specific floor plan is shown in Figure 2.4.

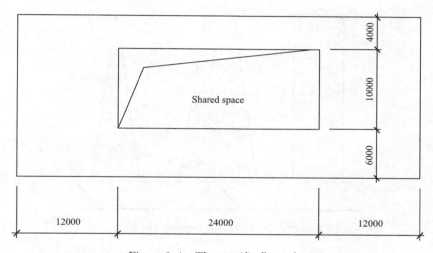

Figure 2.4   The specific floor plan

Question:

Judging its regularity.

Solution:

(1) Floor opening area: $(24\text{m}\times10\text{m})=240\text{m}^2$

Total floor area: $(48\text{m}\times20\text{m})=960\text{m}^2$

30% of the total floor area: $0.3\times960\text{m}^2=288\text{m}^2>240\text{m}^2$

Meeting the requirements of the regular structure on the opening area of the floor.

(2) Effective floor width: $4\text{m}+6\text{m}=10\text{m}$, 50% equal to the width of the floor 20m.

Meeting the requirements of the effective slab width limit of the regular structure.

(3) Minimum net width limit requirements for slabs in any direction: $a_1=4\text{m}>2\text{m}$, $a_2=6\text{m}>2\text{m}$.

$a_1+a_2=10\text{m}>5\text{m}$, and $>0.5L_2=0.5\times20\text{m}=10\text{m}$, meeting the requirements.

2.2.3   Vertical assembly

The complex shape of the building will lead to the uneven distribution of vertical strength and stiffness of the structural system. Under the action of earthquake, a certain layer or a certain part will yield and a large elastic-plastic deformation will occur. For example, a building with a sudden contraction of the façade or a partially protruding building will have a stress concentration at the concave corner.

The building elevation of the seismic zone also requires the use of evenly varying geometries such as rectangles, trapezoids and triangles to avoid the use of stepped elevations with sudden changes (Figure 2.5). Because of the sudden change in the shape of the elevation, it will inevitably lead to drastic changes in mass and lateral stiffness. During an earthquake, the protruding part will be aggravated by intense vibration or concentrated concentration of plastic deformation.

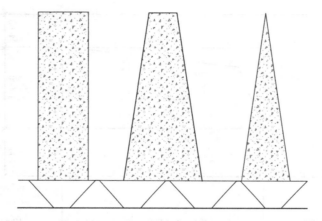

Figure 2.5   Favorable building elevation

The principle of vertical arrangement of the structure, as far as possible, the bearing capacity and vertical stiffness of the structure are gradually reduced from bottom to top, the change is uniform, continuous, and no sudden change occurs. In practical engineering, the cross-sectional dimensions and material strength of the members are often changed along the vertical segment. This changes cause the stiffness to change and should also decrease from bottom to top. From the convenience of construction, the number of changes should not be too much; but from the point of view of structural stress, the number of changes is too small, and each time the change is too large, it is easy to produce a sudden change in stiffness. It is best to reduce the size and strength to stagger the floor to avoid simultaneous changes in the same layer.

Along the vertical stiffness, in addition to the sudden change of the structural stiffness in the vertical direction due to the sudden change of the vertical shape of the building, the structural vertical structure mutation often occurs due to the sudden change of the anti-lateral force structure. If the bottom layer requires a large indoor space, the abrupt change in the stiffness caused by a part of the seismic wall or the frame column is eliminated. At this time, the cross-sectional dimensions of the anti-seismic wall and the lower column should be increased as much as possible, and the concrete grade of these floors should be increased to minimize the degree of stiffness weakening. For example, if the middle floor or the top floor is set to open a large room due to the needs of building function, and some of the seismic walls or the frame columns can be removed, the removed walls should not be too much, and the remaining walls and frame columns should be reinforced to resist the earthquake shear force borne by the removed walls.

The Code for Seismic Design of Buildings GB 50011—2010 gives three types of vertical irregularities, which are shown in Table 2. 7.

Main type of vertical irregularity　　　　　　　　Table 2. 7

Type of irregularity	Definition and reference index
Irregularity of lateral rigidity	The lateral rigidity of this storey is less than 70% of the adjacent upper storey or less than 80% of the average lateral rigidity of the adjacent three storeys(Figure 2. 6);except for the top storey or the small buildings outside roof,the horizontal size of local contraction is larger than 25% of the adjacent lower storey. <div style="text-align:center">Figure 2. 6　Irregularity of lateral rigidity</div>
Discontinuity of vertical lateral-force-resisting components	The internal force of vertical lateral-force-resisting components (columns, seismic walls and seismic bracing) is transmitted downward through horizontal transmission components (beam and truss)
Discontinuity of storey bearing capacity	The inter-storey shear capacity of lateral-force-resisting structure is less than 80% of the adjacent upper storey

The buildings meeting plan irregularity and vertical irregularity shall be adopted with three-dimensional calculation model, the seismicshear force of storeys with small rigidity shall be multiplied by a enhancement coefficient no less than 1. 15.

If one vertical lateral-force-resisting component is discontinuous, the seismic internal force transferred through this component to the horizontal transmission components shall be multiplied by a enhancement coefficient of 1. 25~2. 0 according to the intensity, type of horizontal transmission component, stress condition and physical dimension, etc.

In case of the abrupt discontinuity ofstorey bearing capacity, the shear capacity of the lateral-force-resisting structure at weak storey shall not be less than 65% of that of the adjacent upper storey.

【Example 2. 3】 A six-story office building adopts a cast-in-place reinforced concrete frame structure with a seismic rating of two, of which the beam and column concrete strength grades are C30. The lateral stiffness of each floor of the office building is shown as follows (Table 2.8):

The lateral stiffness of each floor of the office building   Table 2.8

Computing layer	1	2	3	4	5	6
X-direction lateral stiffness(kN/m)	1.0×10^7	1.1×10^7	1.9×10^7	1.9×10^7	1.65×10^7	1.65×10^7
Y-direction lateral stiffness(kN/m)	1.2×10^7	1.0×10^7	1.7×10^7	1.55×10^7	1.35×10^7	1.35×10^7

Question:

Determining the vertical regularity of the structure and the horizontal seismic shear increase factor.

Solution:

First floor: X direction $\dfrac{k_1}{(k_2+k_3+k_4)/3} = \dfrac{1.0}{(1.1+1.9+1.9)/3} = 0.61 < 0.8$

Y direction $\dfrac{k_1}{(k_2+k_3+k_4)/3} = \dfrac{1.2}{(1.1+1.7+1.55)/3} = 0.85 > 0.8$

Second floor: X direction $\dfrac{k_2}{(k_3+k_4+k_5)/3} = \dfrac{1.1}{(1.9+1.9+1.65)/3} = 0.61 < 0.8$ or

$\dfrac{k_2}{k_3} = \dfrac{1.1}{1.9} = 0.58 < 0.7$

Y direction $\dfrac{k_2}{(k_3+k_4+k_5)/3} = \dfrac{1.0}{(1.7+1.55+1.35)/3} = 0.65 < 0.8$ or $\dfrac{k_2}{k_3} = \dfrac{1.0}{1.7} = $

$0.59 < 0.7$

Therefore, they are all irregular types of vertical stiffness.

2.3 Seismic joints

2.3.1 Arrangement requirements

For the buildings with complex configuration and irregular plan and vertical planes, a comparative analysis of such factors as degree of irregularity, condition of base foundation and technical economy shall be conducted to determine whether set seismic joints, and the following requirements shall be met respectively:

(1) If no seismic joint is set, practical calculation model shall be adopted to analyze and distinguish the vulnerable due to stress concentration, deformation concentrated or earthquake twisting effect, so as to adopt corresponding strengthening measure.

(2) For the buildings with complex configuration and irregular plan and vertical planes, a comparative analysis of such factors as degree of irregularity, condition of base foundation and technical economy shall be conducted to determine whether set seismic joints, and the following requirements shall be met respectively.

(3) To set expansion joint and settlement joint, their width shall comply with the re-

quirements on seismic joint.

Reasonable installation of seismic joints can divide a complex building into a "regular" structural unit. as the picture shows (Figure 2.7), the irregularly L-shaped building is divided into two regular rectangular structural units by a seismic joints. The installation of seismic joints can reduce the difficulty of structural seismic design and improve the seismic performance of each structural unit, but it also brings many new problems. For example, because the wall or frame column is required on both sides of the seismic joint to make the structure complicated, especially making the foundation processing more difficult, and the building is inconvenient to use, and the building elevation is difficult to handle. What is more prominent is that the structures on both sides of the seam enter an elastic-plastic state during the earthquake, and the displacement increases sharply and collides with each other, causing serious earthquake damage such as light exterior decoration, parapet wall, cornice damage, serious body structure damage. Therefore, complex buildings do not always advocate the installation of seismic joints. The plane size and structural arrangement should be adjusted and structural or construction measures should be taken. When there is no seismic joint, seismic separation shall be carried out and structural measures to enhance ductility shall be adopted. If no measures are taken or if seismic joints must be provided, the necessary seismic joint width must be ensured to prevent damage.

Figure 2.7　Seismic joint setting

In the following cases, the seismic joints should be set up to divide the whole building into several independent structural units of regulars:

(1) Plane shapes are irregular or vertical are irregular.

(2) The length of the house exceeds the maximum spacing of the construction and expansion joints specified in the relevant Structural Design Code, and there is no condition to take special measures and the expansion joint must be set.

(3) The soil quality of the foundation is uneven or the load of the superstructure is quite different. The expected settlement of each part of the house is too large, and the settlement joint must be set.

(4) The structural system of each part of the house is completely different, and the quality or lateral displacement stiffness is very different.

【Example 2.4】 A frame shear wall structure, seismic precautionary intensity is 8, its building plane as shown in Figure 2.8. It is planned to set up four seismic joints ①, ②, ③, ④.

Question:

Which seismic joints must be set?

Solution:

According to the *Code for Seismic Design of Buildings* GB 50011—2010,

seismic joints ①: $\dfrac{B}{B_{max}}=\dfrac{3}{(3+16+2)}=0.14<0.3$ Don't have to set.

seismic joints ②, seismic joints ③: Don't have to set.

seismic joints ④: $\dfrac{B}{B_{max}}=\dfrac{15}{(15+16+3)}=0.44>0.3$ Must be set.

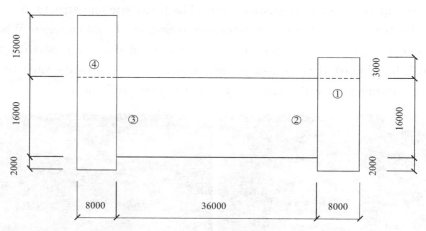

Figure 2.8   Layout plan

2.3.2   Calculation width of seismic joint

When setting the seismic joints, it should meet the following requirements:

(1) The seismic joints width should meet the following requirements:

1) For frame structures, the height should not be less than 100mm when the height is less than 15m; when it is more than 15m, the height of each of 6 degrees, 7 degrees, 8 degrees and 9 degrees is 5m, 4m, 3m, and 2m, respectively, and should be widened by 20mm.

2) The value of seismic joints width of frame-shear wall structure houses shall not be less than 70% of the values specified in item 1) of this paragraph. The value of seismic joints width of the shear wall structure house shall not be less than 50% of the value specified in item 1) of this paragraph, and neither shall be less than 100mm.

(2) When the structural systems on both sides of the seismic joints are different, the

seismic joints width should be determined according to the unfavorable structure type.

(3) When the height of the houses on both sides of the seismic joints is different, the width of the seismic joints can be determined according to the lower height.

(4) Seismic design of the frame structure housing under 8 or 9 degree, when the height of the structural layers on both sides of the anti-seismic joint is large, the stirrups of the frame columns on both sides of the anti-seismic joint should be encrypted along the full height of the house, and can be placed along the full height of the house as needed. Each setting shall be no less than two anti-collision walls perpendicular to the anti-seismic joint.

(5) When there is a large settlement difference in the foundation of the adjacent structure, it is advisable to increase the width of the seismic joints.

(6) The seismic joints should be set along the full height of the house. The basement and foundation may not have seismic joints, but the structure and connection should be strengthened at the corresponding position with the upper seismic joints.

(7) It is not advisable to use a cow-leg joist between structural units or between the main building and the podium to provide seismic joints, otherwise reliable measures should be taken.

【Example 2. 5】 In the 7 degree seismic precautionary zone, a frame-shear wall structure building with a height of 60m (distance from the outdoor floor to the roof). There is also a 4. 50m elevator room on the roof. The other adjacent building is a frame structure hall with a height of 20m. The outdoor elevation difference on both floors is 0. 60m.

Question:

Determining its seismic joints width.

Solution:

When the structural systems on both sides of the seismic joint are different, the seismic joint width shall be determined according to the unfavorable structure type, and determined according to the lower house height. In this question, the seismic joint width should be determined according to the frame structure of 20m. The precautionary intensity is 7 degree. Seismic joint width is:

$$\delta = 100 + \frac{(20+0.6)}{4} \times 20 = 128 \text{mm}$$

【Example 2. 6】 In a large urban area with a intensity of 7 degree, a high-rise commercial and residential complex is proposed. Among them, the commercial building adopts a frame-shear structure with a height of 30m; the residential building has a height of 50m and adopts a shear wall structure; the basement of the two buildings does not have a settlement joint, and the upper structure is divided into independent seismic units by seismic joints.

Question:

Find the minimum width of the seismic joint.

Solution:

The seismic joint width is determined by the lower frame shear structure: $H=30$m. Frame structure:

$$\delta=100+\frac{(30-15)}{4}\times 20=175\text{mm}$$

Frame shear wall structure: $\delta=0.7\times175=123$mm

2.4 Structural system

The structural system shall be determined by comprehensive comparison of technical, economic and use conditions according to the seismic precautionary category, seismic precautionary intensity, building height, site conditions, foundation, structural materials and construction factors.

The seismic structure should adopt a reasonable and economical structure type, and whether the structural scheme is reasonable or not plays a major role in safety and economy. The seismic response of the structure is closely related to the site characteristics. The ground motion characteristics of the site are related to the earthquake focal mechanism, magnitude and epicentral distance. The importance of the building and the level of decoration have limitations on the lateral deformation of the structure. Consideration should also be given to constraints on structural materials and construction conditions, as well as licensing of economic conditions.

2.4.1 Basic requirements of the structural system

(1) The structural system should have a clear calculation sketch and a reasonable seismic action propagation route. The seismic structural system requires clear force, reasonable force transmission and uninterrupted force transmission, the seismic analysis of the structure is more in line with the actual performance of the structure during the earthquake and is very beneficial to improve the seismic performance of the structure, which is one of the factors considered in the selection of structure and the arrangement of lateral-force-resisting system.

(2) The loss of seismic capacity or load carrying capacity against gravity loads of the entire structure due to damage to some structures or components should be avoided. Studies of macroscopic phenomenon of collapse of earthquakes have shown that the most direct cause of collapse of a house is the ability of the structure to lose its ability to withstand gravity loads due to damage. Therefore, in any case, the carrying capacity of the structure to the gravity load should be ensured first.

(3) The structure should have the necessary seismic capacity, good deformation capacity and the ability to consume seismic energy. Among them, "necessary seismic capacity" means that the structure should have the necessary strength, and "good deformability" means that the deformation of the structure does not cause structural function loss or ex-

ceeds the allowable damage, "good ability to consume seismic energy" refers to the ability of a structure to absorb and consume the energy of a seismic input, that is, good ductility. The seismic capacity of the structure requires uniformity of strength, stiffness and deformation capacity, that is, the seismic structural system should have the necessary strength and good deformation energy consumption. When there is only strength and lacks sufficient ductility, it is easily destroyed under strong earthquakes. Although it has good ductility and insufficient strength, it will inevitably produce great deformation under strong earthquakes, and the damage will be severe or even collapse.

(4) Measures should be taken to improve the seismic resistance of weak parts that may occur.

2.4.2   Other requirements of the structural system

The structure should have multiple anti-seismic defense lines. The earthquake damage survey shows that the destructive strong earthquake has the characteristics of long duration and many times of pulse reciprocation. Once the house of a single institutional system is destroyed, the continuous vibrations that follow will cause the house to collapse. When the house adopts multiple lines of defense, after the first line of defense is destroyed, the subsequent line of defense can replace the subsequent ground motion shock, thereby ensuring the minimum safety of the house and avoiding the collapse of the house. Therefore, it is necessary to set up multiple defense lines for earthquake-resistant houses, and it is also the basic requirement for "big earthquakes do not fall". An earthquake-resistant structural system shall consist of several sub-systems with better ductility and shall be connected by structural members with good ductility, such as:

(1) The frame-shear wall system consists of two systems: a ductile frame and an anti-seismic wall. Among them, in the frame-shear wall structure, the earthquake resistance is the first line of defense against earthquakes due to its large lateral stiffness, and the frame is the second line of defense against earthquakes. In the frame structure with few seismic walls, because the number of seismic walls is small, it cannot be a line of defense, and the structural system does not belong to the structural system of multiple lines of defense.

(2) The double-limb wall or multi-limb seismic wall system consists of several single-limb wall systems. When the earthquake occurs, the coupling beam first yields and absorbs a large amount of seismic energy, which can transmit bending moments and shear forces, and can also have certain constraints on the wall limbs.

(3) In the vertical system of a single-storey building, the inter-column support is the first line of defense, and the column is the second line of defense, and the structural energy is ensured by the yield energy of the support between the columns.

(4) For the ductile frame, the frame beam belongs to the first line of defense, and the deformation of the beam consumes energy, and the yield is prior to the frame column

so that the column is in the second line of defense.

Earthquake damage surveys show that the collapse of the house is due to the ability of the lateral force-resisting members to lose vertical loads. Therefore, the structure should have a reasonable distribution of stiffness and bearing capacity, to avoid weakening or weakening due to local weakening or sudden formation, resulting in excessive stress concentration or plastic deformation concentration. The sudden increase or decrease in the lateral stiffness of the floor is a sudden change in stiffness. The sudden change in stiffness is caused by the complexity of the building or the discontinuity of the main lateral force system in the vertical arrangement. The location of stress abrupt change will also result in stress concentration and deformation concentration. If the stress concentration is not properly strengthened, it will enter the plastic deformation stage before the adjacent part, causing plastic deformation to concentrate, eventually leading to serious damage or even collapse.

2.5　Materials and construction of the structure

2.5.1　Structural material performance index

1. The materials of masonry structure shall meet the following requirements:

(1) The strength grade of common brick and perforated brick shall not be less than MU10 and the strength grade of their masonry shall not be less than M5.

(2) The strength grade of small concrete hollow block shall not be less than MU7.5, and the strength grade of their masonry shall not be less than Mb7.5.

2. The materials of concrete structure shall meet the following requirements:

(1) The strength grades of concrete for frame-supported beams and columns as well as frame-supported beams and columns and node-core area assigned to seismic Grade Ⅰ shall not be less than C30; and the strength grades of concrete for constructional columns, core columns, ring-beams and other components shall not be less than C20.

(2) For the frames and diagonal bracing components (including the stair section) assigned to seismic Grade Ⅰ, Ⅱ and Ⅲ, if ordinary reinforcements are used as their longitudinal bearing force reinforcements, then the ratio between the measured tensile strength and the measured yield strength of steel reinforcement shall not be less than 1.25, the ratio between the measured value and standard value of yield strength shall not be larger than 1.3, and the measured overall elongation of steel reinforcement under the maximum tensile stress shall not be less than 9%.

3. The steels of steel structures shall meet the following requirements:

(1) The ratio between the measured yield strength and measured tensile strength of steels shall not be larger than 0.85.

(2) The steels shall have obvious yield steps and their elongation rate shall not be less than 20%.

(3) The steels shall have good weld ability and qualified impact ductility.

4. The performance indexes of structural materials still should meet the following requirements:

(1) The ordinary reinforcements with better ductility, tenacity and weld ability should be preferred; for the strength grade of ordinary reinforcements, the longitudinal bearing force reinforcements should be selected the hot rolled reinforcements with seismic performance index no less than Grade HRB400 or may be adopted the Grade HRB335 hot rolled reinforcements; the stirrups should be selected the hot rolled reinforcements with seismic performance index no less than Grade HRB335 or may be adopted the Grade HPB300 hot rolled reinforcements.

(2) The concrete strength grade of concretestructures, like seismic wall, should not exceed C60 for intensity 9 and C70 for intensity 8 for other components.

(3) The steel type of steel structures should be selected the Grade Q235-B, C, D carbon structural steels and Grade Q345-B, C, D and E high strength low alloy structural steels; when reliable references are available, other types and grades of steels also may be adopted.

2.5.2　Earthquake structure requirements for construction quality

1. steel bar replacement

In the construction, when it is required to replace the longitudinally stressed steel bars in the original design with the steel bars with higher strength grades, the requirements of the strength conditions cannot be emphasized on one side, and the ductility of the structure is also ensured. In the seismic design, not more steel bars are better. If more concrete is destroyed first, it is brittle failure, which will reduce the deformation ability of the components. It should be converted according to the principle that the design values of the tensile load capacity of the steel bars are equal (this is the "equal strength substitution" $f_{y1}A_{s1} = f_{y2}A_{s2}$). Because the total yield strength of the main force after replacement is not higher than the total yield strength of the original design of the main section of the section, it can avoid the transfer of weak parts and the brittle failure of the components in the affected parts (such as coagulation crushing, reinforced concrete component shear failure, etc.). It should also be noted that the structural requirements such as the deflection and crack width of the normal use stage and the minimum reinforcement ratio and the spacing of the steel bars after the change of the strength and diameter of the steel bar should meet the requirements of the normal use limit state and seismic construction measures.

【**Example 2.7**】 A supermarket is a four-story reinforced concrete frame structure with a building area of $25,000m^2$, this building height of 24m, and a seismic precautionary intensity of 7 degree. The longitudinal reinforcement of the original design of the frame column is 8 Φ 22. During the construction process, due to the supply of raw materials on site, it is proposed to replace the steel bars in Table 2.9.

Calculation data table   Table 2. 9

Reinforcement	The measured yield strength σ_S (MPa)	The measured tensile strength σ_b (MPa)
20	438	550
25	370	510
20	492	610

（A）8Φ20

（B）4Φ25（Corner）+4Φ20（Central）

（C）8Φ25

（D）4Φ25（Corner）+4Φ20（Central）

Question:

Which of the following is most appropriate?

Solution:

（B）

Take seismic measures. According to the *Code for Seismic Design of Buildings* GB 50011—2010, the seismic rating of the frame shall be Grade Ⅱ.

According to the *Code for Seismic Design of Buildings* GB 50011—2010, the ratio between the measured tensile strength and the measured yield strength of steel reinforcement shall not be less than 1.25, the ratio between the measured value and standard value of yield strength shall not be larger than 1.3, see the Table 2.10 below for related calculations.

Calculation table   Table 2. 10

Reinforcement	The measured tensile strength / the measured yield strength	The measured value of yield strength / the standard value of yield strength
20	1.256>1.25（Y）	1.307>1.3（N）
25	1.378>1.25（Y）	1.1<1.3（Y）
20	1.24<1.25（N）	1.47>1.3（Y）

So, in the Table 2.10, Reinforcement does not meet the specification requirements, it cannot be used for secondary frame columns, answer （B）.

2. Horizontal construction joint

The horizontal construction joints of concrete wall and frame column shall be taken with measures to strengthen the bonding property of concrete, For the seismic Grade Ⅰ wall and the connection part between transition storey slab and ground concrete wall, the shear bearing capacity of the section of horizontal construction joint shall be checked and calculated.

Exercises

2.1　What is the basic principle of plane and elevation layout of buildings?

2.2　Why should the ratio ofstructural height to width be controlled?

2.3　What is the "multiple seismic defense lines"?

2. 4　What is the relationship among the structural stiffness, the bearing capacity of structure and the structure ductility?

2. 5　What is the effect of site conditions on the seismic building? How to consider the effect in the seismic design of building?

2. 6　How to consider the seismic joint in the seismic design?

2. 7　Why do the seismic buildings have the requirements of the concrete grade and the reinforcement performance?

Chapter 3 Site, Base and Foundation

The construction site refers to the location of the building, which is roughly equivalent to the factory area, the residential area and the natural village on the plane. All of the buildings are located and embedded on the rock foundation of the construction site. The damage to the building caused by the earthquake is transmitted to the superstructure through the site, base and foundation. Therefore, studying the seismic damage modes, failure mechanisms and seismic design of building structures under earthquakes is inseparable from the study of site soil and base. The reaction of the research site and foundation under earthquake action and its influence on the superstructure is an important task for seismic evaluation of the site.

After a strong earthquake, a large number of buildings and structures will be damaged. The damage forms are various and extremely complex, but we can roughly distinguish them into two types from the perspective of the nature of the damage and engineering countermeasures, which are the destruction of the site and base and the ground motion of the site.

3.1 Site

3.1.1 Site and seismic effect of site

The site is the area of the building group, the scope of site is roughly equivalent to the factory area, residential area and natural villages, it generally refers to the land not less than $1km^2$. The earthquake response of the site to the building is straightforward, the quality of the site conditions will directly affect the size of the earthquake action of the building.

When choosing a building site, a comprehensive evaluation should be made on the favorable, common, unfavorable and hazardous sections according to the engineering needs and relevant data of seismic activity, engineering geology and seismological geology. For the unfavorable areas, avoid requirements should be put forward; when it is unable to avoid, effective measures should be taken. For hazardous areas, it is strictly prohibited to build buildings of precautionary category A and B, and category C buildings should not be built. The construction site shall be divided into sections that are favorable, general, unfavorable and dangerous to the building in accordance with Table 3.1.

Division of favorable, common, unfavorable and hazardous sections   Table 3.1

Section type	Geological, topographical, geomorphic description
Favorable section	Stable rock bed, stiff soil, or wide-open, even, compacted and homogeneous medium-stiff soil
Common section	Sections not belonging to the favorable, unfavorable and hazardous sections

Continued

Section type	Geological，topographical，geomorphic description
Unfavorable section	Soft soil，liquefied soil，stripe-protruding spur，lonely tall hill，steep slopes，steep step，river bank or boundary of side slops，soil layer having obviously heterogeneous cause of formation，rock character and state in plane(including abandoned river beds，loosened fracture zone of fault，and hidden swamp，creek，ditch and pit，as well as base formatted with excavated and filled)，plastic loess with high moisture，ground surface with structural fissure，etc.
Hazardous section	Places where landslide，collapse，land subsidence，ground fissure and debris flow may occur during the earthquake，as well as the positions in causative fault where ground dislocation may occur

Earthquake damage shows that the influence of local topographic conditions on seismic intensity is closely related to the composition of geology，and the influence of non-rock topography on seismic intensity is more obvious than that of rock topography. Due to the structure design in the construction site selection is generally accepted，therefore，in the preliminary design stage，the site should be paid special attention to the construction site to judge again. Construction sites should not be selected in dangerous areas. For unfavorable areas，corresponding technical measures should be taken according to the unfavorable degree.

From the analysis on the principle of seismic wave propagation in rock，it already has a plurality of frequency components，which in the period of amplitude spectrum of frequency components in the maximum amplitude of the corresponding，known as the predominant period of ground motion. In the process of seismic wave propagating to the surface through the covering soil，some frequency wave groups consistent with the natural period of the soil layer will be amplified，while others will be attenuated or even completely filtered out. In this way，the predominant period of ground motion depends to a great extent on the inherent period of the site due to the filtering characteristics and selective amplification of the soil layer after the seismic wave through the soil layer. When the natural period of a building is close to the predominant period of ground motion，the vibration of the building will increase，and accordingly，the earthquake damage will be aggravated.

Further theoretical analysis proved that the seismic effect of multi-layer soil mainly depends on the three basic factors：soil covering thickness，shear wave velocity and rock impedance ratio. Among the three factors，the rock impedance ratio mainly affects the resonance amplification effect，while the others mainly affect the frequency characteristics of ground motions.

3. 1. 2　The cover layer thickness of the site

The cover layer thickness of the site is originally defined as the distance from the ground surface to the underlying bedrock surface. From the point of view of seismic wave propagation，the bedrock interface is a strong refraction and reflection surface in the seismic wave propagation path，and the vibration stiffness of the strata below this interface is much larger than the corresponding value of the upper soil layer. According to this background, it is often judged that when the shear wave velocity of the lower soil layer is 2.5 times that of the upper soil layer, and there is no rock and soil layer whose shear wave velocity is less than 400m/s in the lower soil layer, the lower soil layer can be regarded as the bedrock approximately. Because it is difficult to obtain the shear wave velocity data of deep soil layer by means of engineering geological investigation, in order to beconvenient on practical, the absolute stiffness of soil layer is further adopted to define the cover layer thickness in the *Code for Seismic Design of Buildings* GB 50011—2010, that is, the shear wave velocity of underground bedrock or its underlying soil layer is greater than 500m/s (and the shear wave velocity of its underlying soil layer is not less than 500m/s). The distance between the hard soil and the surface is known as "the cover layer thickness of the site".

The cover layer thickness at the building site shall be determined according to the following requirements:

(1) Generally, this thickness shall be determined according to the distance from the ground surface to the top surface of a soil layer, under which the shear wave velocity is more than 500m/s and the shear wave velocity of the soil layers under it is not less than 500m/s.

(2) If such soil layer with shear wave velocity is more than 2.5 times of that of the soil layers above it exists 5m under the ground surface and the shear wave velocity of this soil layer and those under it all is less than 400m/s, then the cover layer thickness may be determined according to the distance from the ground surface to the top surface of this soil layer.

(3) The lone-stone and lentoid-soil with shear-wave velocity greater than 500m/s shall be deemed the same as the surrounding soil layer.

(4) The hard volcanic inter-bedded rock in the soil layer shall be deemed as rigid body and its thickness shall be deducted from the thickness of cover soil layer.

3.1.3 The site class of building

The previous analysis mentioned that there are obvious differences in the frequency spectrum of ground motion on different sites. In order to show this characteristic, the building site is dividied into four categories in the *Code for Seismic Design of Buildings* GB 50011—2010, in which the first category is divided into two sub-categories: I_0 and I_1. If the reliable shear wave velocity and the cover layer thickness are available and their values are near to the divisional line of site class listed in Table 3.2, then it shall be allowed to determine the characteristic period for seismic action calculation with interpolation method.

As can be seen from Table 3.2, the site class is determined by the equivalent shear wave velocity of soil layer and the cover layer thickness of the site.

Cover layer thickness (m) of various building sites　　　　　Table 3. 2

Shear wave velocity of rock or equivalent shear wave velocity of soil (m/s)	Site class				
	I_0	I_1	II	III	IV
$v_s > 800$	0				
$800 \geqslant v_s > 500$		0			
$500 \geqslant v_{se} > 250$		< 5	$\geqslant 5$		
$250 \geqslant v_{se} > 150$		< 3	$3 \sim 50$	> 50	
$v_{se} \leqslant 150$		< 3	$3 \sim 15$	$15 \sim 80$	> 80

Note: v_s refers to the shear wave velocity of rock.

The classification standard of Table 3. 2 is mainly applicable to the general situation of shear wave velocity increasing with depth. In practical engineering, the influence of layered soil interlayer is complicated, so it is hard to use a single indicator to describe it. The analysis of seismic response show that the influence of hard soil interlayer is relatively small, although the soft soil interlayer with deep burial depth and large thickness can inhibit the high-frequency component of the input seismic wave of bedrock, but it can significantly amplify the low-frequency component of the input seismic wave. Therefore, when there is obvious soft soil interlayer below the depth of calculation, the site class can be appropriately increased.

The measurement of the shear wave velocity of soil layer shall meet the following requirements:

(1) At the stage of primary investigation of the site, for the same geologic units in large area, the number of drilling holes for testing the shear wave velocity of soil layer should not be less than 3.

(2) At the stage of detailed investigation of the site, for every building, the number of drilling holes for testing the shear wave velocity of soil layer should not be less than 2; if the data varies significantly, the number may be increased properly. For the crowded building complex in one community, which are built in the same geological unit, such number may be reduced properly but that for each tall building and each large-span spatial srrucn1re shall not be less than 1.

(3) For buildings assigned to Category D or to Category C with no more than 10 storeys and no more than 30m in height, if the measured shear wave velocity is not available, the shear wave velocity of each soil layer may be estimated within the range of shear wave velocity specified in Table 3. 3 based on the name and character of rock-soil and according to the soil types listed in Table 3. 3 and the local experience.

Classification of Soil Types and Scope of Shear Wave Velocity　　　　　Table 3. 3

Soil type	Name and character of rock-soil	Scope of shear wave velocity of soil layer(m/s)
Rock	Stiff, hard, and complete rocks	$v_s > 800$

Continued

Soil type	Name and character of rock-soil	Scope of shear wave velocity of soil layer(m/s)
Stiff soil or soft sock	Broken and comparatively broken rock, soft and comparatively soft rock, compact gravel soil	$800 \geqslant v_s > 500$
Medium-stiff soil	Medium dense or slightly dense detritus, dense or medium-dense gravel, coarse or medium sands, cohesive soil and silt with $f_{ak} > 500$, hard loess	$500 \geqslant v_s > 250$
Medium-soft soil	Slightly dense gravel, coarse and medium sands, fine and mealy sands (excluding the loose sand), cohesive soil and silt with $f_{ak} \leqslant 150$, filled soil with $f_{ak} > 130$kPa, plastic young loess	$250 \geqslant v_s > 150$
Soft soil	Mud and muddy soil, loose sand, new sedimented cohesive soil and silt, filled soil with $f_{ak} \leqslant 130$, flow plastic loess	$v_s \leqslant 150$

Note: f_{ak} is the characteristic value (kPa) of base bearing capacity obtained through load test or other methods; v_S is the shear wave velocity of rock-soil.

The equivalent shear wave velocity of soil layer shall be calculated according to the following formulas:

$$v_{se} = d_0 / t \tag{3-1}$$

$$t = \sum_{i=1}^{n} (d_i / v_{si}) \tag{3-2}$$

Where v_{se} ——equivalent shear wave velocity of soil layer (m/s);

d_0 ——calculation depth (m), which shall be taken as the smaller value between the cover layer thickness and 20m;

t ——travel time of shear wave between the ground and the calculation depth;

d_i ——thickness (m) of the ith soil layer within the range of calculation depth;

v_{se} ——shear wave velocity (m/s) of the ith soil layer within the range of calculation depth;

n ——number of soil layers within the range of calculation depth.

【Example 3. 1 】

The construction site drilling known geological data is shown in Table 3. 4, determining the site category.

Borehole data   Table 3. 4

Bottom depth of soil layer(m)	Thickness of soil layer(m)	Name of rock and soil	Shear wave velocity of soil layer(m/s)
1. 5	1. 5	Miscellaneous filled soil	180
3. 5	2. 0	Silty soil	240
7. 5	4. 0	Fine sand	310
12. 5	5. 0	Medium sand	520
15. 0	2. 5	Gravel sand	560

Solution:

(1) Determination of the cover layer thickness of the site.

Because the soil below 7.5m is below the surface, $v_s = 520$m/s>500m/s and lower lying layer $v_s>500$m/s. So $d_0 = 7.5$m.

(2) Calculate the equivalent shear wave velocity, according to formula:

$$v_{se} = 7.5 / \left(\frac{1.5}{180} + \frac{2.0}{240} + \frac{4.0}{310} \right) = 253.6 \text{m/s}$$

As the Table 3.2 shows, v_{se} located between $200 \sim 500$m/s, and $d_0 > 5$m, So it belongs to category II site.

【Example 3.2 】

The equivalent shear wave velocity of the site is calculated according to Table 3.5, and determine the site category.

Shear wave velocity of soil layer Table 3.5

Thickness of soil layer	2.2	5.8	8.2	4.5	4.3
v_s (m/s)	180	200	260	420	530

Wrong solution:

(1) Determination of the cover layer thickness of the site.

Because the soil below 20.7m is below the surface, $v_s = 530$m/s$ > 500$m/s, so $d_0 = 20.7$m.

(2) Calculate the equivalent shear wave velocity, according to formula:

$$v_{se} = 20.7 / \left(\frac{2.2}{180} + \frac{5.8}{200} + \frac{8.2}{260} + \frac{4.5}{420} \right) = 247.9 \text{m/s}$$

As the Table 3.2 shows, v_{se} located between $140 \sim 250$m/s, and $d_0 > 3$m, so it belongs to category II site.

Right solution:

(1) Determination of the cover layer thickness of the site.

Because the soil below 20.7m is below the surface, $v_s = 530$m/s$ > 500$m/s, So $d_0 = 20$m.

(2) Calculate the equivalent shear wave velocity, according to formula:

$$v_{se} = 20 / \left(\frac{2.2}{180} + \frac{5.8}{200} + \frac{8.2}{260} + \frac{3.8}{420} \right) = 244.5 \text{m/s}$$

As the Table 3.2 shows, v_{se} located between $140 \sim 250$m/s, and $d_0 > 3$m, so it belongs to category II site.

3.1.4　Site zoning

For cities of medium size or above, the code for seismic design of buildings allows the use of approved seismic zoning for seismic fortification. It involves the area division of site design ground motion. In general, the results of site classification, ground motion parameters and ground failure potential zoning in urban area are given. Next, the article simply introduce the basic content of site zoning.

The basic method and process of site zoning are:

(1) Collect engineering geology, hydrogeology, seismological and geological data within the urban area.

(2) Based on the above-mentioned materials, the control geological profile of the area under consideration is drawn, and the plane control points of site zoning are established.

(3) Conduct supplementary geological prospecting and shear wave velocity tests as appropriate.

(4) According to the statistical data of engineering geology, the empirical relationship between shear wave velocity and depth of different types of soils is given.

(5) The equivalent shear wave velocity of shallow rock and soil (20m below the surface) at each control point is calculated according to the control geological profile and the empirical relationship of shear wave velocity, and the thickness of each control point is determined.

(6) According to the equivalent shear wave velocity and the thickness of the overburden, the site within the urban area is zoned according to the provisions of Table 3. 2.

The equivalent shear wave velocity contours and the inherent periodic contours of the site can also be obtained by in-depth site regionalization. The T of the site can be calculated according to the shear wave repeated reflection theory.

$$T = \sum_{i=1}^{n} \frac{4d_i}{v_{si}} \tag{3-3}$$

The middle symbol indicates the same formula (3-1) and (3-2).

Meticulous site zoning can save investment once and for all. Architectural seismic designers should consult the local competent seismic department for relevant information and apply it to the design according to specific conditions.

3. 1. 5 Building site evaluation

The *Code for Seismic Design of Buildings* GB 50011—2010 stipulates that when there is an earthquake fracture in the site, the impact of the fault on the project shall be evaluated and the following requirements shall be met:

(1) For the cases that meet the following requirements, the impact of earthquake rupture and dislocation on the ground buildings can be ignored.

1) The seismic fortification intensity is less than intensity 8;

2) Non Holocene active faults;

3) When the seismic fortification intensity is intensity 8 or intensity 9, the thickness of the soil cover layer of the buried fault is greater than 60m and 90m respectively.

(2) In the event that the situation does not conform to the provisions in Item 1 of this article, the main fault zone shall be avoided in the selection of site. The avoidance distance

should not be less than the minimum avoidance distance for causative fault specified in Table 3. 6. If scattered Category C and D buildings with less than 3 storeys are required to be built within the scope of avoidance distance, the seismic measures of one intensity higher shall be taken, the integrity of foundation and superstructure shall be improved, and the building must not cross over any fault trace.

Minimum avoidance distance of causative fault (m)　　　　　Table 3. 6

Earthquake intensity	Types of seismic fortification for buildings			
	A	B	C	D
8	Special study	200m	100m	—
9	Special study	400m	200m	—

The code for seismic resistance stipulates that when buildings of Class C and above are required to be constructed in unfavorable sections such as strip-shaped mouths, high isolated hills, steep slopes of non-rock and strongly weathered rocks, river rocks and edges of slopes, the stability of buildings of Class C and above shall be ensured in addition to the stability under seismic action, the parameters of design ground motion in unfavorable sections shall also be estimated. The maximum horizontal seismic influence factor multiplied by the magnification factor should be determined according to the specific conditions of the unfavorable sections and used in the range of 1. 1~1. 6.

Site geotechnical investigation should be based on the actual needs of the division of favorable, general, unfavorable and dangerous sections of the building, to provide building site types and rock and soil seismic stability (such as landslide, collapse, liquefaction and seismic subsidence characteristics) evaluation, for the need to use time-history analysis method to supplement the calculation of the building, should also be based on design requirements provide soil profile, thickness of site cover layer and related dynamic parameters.

3. 1. 6　Ground motion characteristics during earthquake

Seismic waves are complex waves. According to the principle of harmonic analysis, it can be seen as a superposition of N simple harmonic waves. Site soil can amplify all kinds of harmonic components transmitted from bedrock, but some of them amplify much more and others less. That is, different site soils have different magnification effects on seismic waves. Understanding the effect of site on seismic waves is of great significance to seismic design and damage analysis of buildings.

From the introduction of seismic waves in Chapter 1, the vibration process at any point on the ground actually includes the comprehensive action of various types of seismic waves. Therefore, the most obvious feature of the ground motion record is its irregularity. From the perspective of engineering application, limited elements to reflect irregular seismic waves can be used. For example, the maximum amplitude can quantitatively reflect the intensity characteristics of ground motions; the periodic distribution characteristics of ground motions can be revealed by analyzing the frequency spectrum of ground motion records; and the degree of cyclic action of ground motions can be investigated by defining and measuring the duration of strong earthquakes.

The acceleration record of ground motion during earthquake is the basic data of earthquake engineering. Acceleration records (time history curves) of strong ground motion are used in drawing spectrum curves of acceleration response and in direct dynamic calculation of structural seismic response. Strong ground motion can be measured by strong motion seismograph. The accelerometer can measure the acceleration time history curve at the location. At present, the vast majority of strong seismographs record only two horizontal and one vertical ground time history curves of the observation points.

What physical quantities are used to describe a strong motion ground motion? It is generally considered that three characteristic parameters can be used to express peak acceleration, duration and main period. Generally speaking, when the magnitude is large, the peak acceleration is high and the duration is long, while the main period varies with the site type and epicenter distance. The larger the site category, the farther the epicentral distance is, the longer the main period (or characteristic cycle) of the earthquake.

The relationship between the components of ground acceleration of strong earthquakes is roughly proportional to statistics. From most seismic records, the average intensity of the two horizontal components of ground motion is approximately the same, and the vertical component of ground motion is equivalent to $1/3 \sim 2/3$ of the horizontal component.

3.2　Natural base and foundation

3.2.1　Principles of foundation seismic design

The base refers to the soil layer within the foundation of the building under the stress level. The statistical analysis of historical earthquake damage data shows that there are few problems in the general land base during earthquakes. The main damage to the upper buildings is the soft soil foundation and the uneven ground. Therefore, the design of the earthquake area of the building should use different treatment plan according to the different conditions of soil.

1. Soft land base

In the earthquake area, the saturated silt and silt soil, the filling soil and the miscellaneous fill soil, the uneven foundation soil, cannot be treated directly to use as the natural base of the building. The engineering practice has proved that although these foundation soils have certain bearing capacity under static conditions, in the earthquake, the load capacity will be lost in all or part or uneven subsidence and excessive subsidence may occur due to the influence of the ground motion, causing the damage or the influence of the building to normal use. The failure of soft soil foundation can not be overcome by widening foundation and strengthening superstructure, so the ground treatment measures (such as displacement, encryption, dynamic compaction, etc.) should be used to eliminate the dynamical instability of soil, or the deep foundation such as pile foundation can avoid the unfavorable influence of the base to the upper building.

2. General land base

The statistic data of the earthquake damage survey show that the buildings built on the general soil natural base are rarely damaged by the lack of base strength or the large subsidence caused by the earthquake. Therefore, according to the *Codes for Seismic Design of Buildings* GB 50011—2010. For the following buildings, the seismic bearing capacity check of natural base and foundation may not be carried out:

(1) Ordinary single-storey factory buildings and single-storey spacious buildings;

(2) Masonry buildings;

(3) Ordinary civil framed buildings and buildings with frame-seismic wall not exceeding 8 storeys and 24m in height.

Multi-storey frame factory buildings and multi-storey buildings with concrete seismic wall, of which the foundation load is equivalent to those specified in Item (3).

Note: The soft cohesive soil layer refers to the soil layer with characteristic value of base bearing capacity less than 80, 100 and 120 respectively for Intensity 7, 8 and 9.

3. Prevention and control of ground fissure hazards

When the earthquake intensity is more than Intensity 7, the ground fissure is easy to develop in the soft site soil and the middle soft site soil area, the buildings, especially brick structures, are often torn apart by ground cracks. Therefore, for the building in the weak site, when the basic intensity is more than Intensity 7, should take measures to prevent ground crack. For example, for a brick-structure house, a cast-in-situ reinforced concrete ring beam can be set up on the basis of a load-bearing brick wall; for single-storey reinforced concrete column factory building, cast-in-place integral foundation brick wall beams or integral foundation wall beams with cast-in-place joints can be set up along one circle of external walls. When the basic intensity is Intensity 9, the above anti-ground crack measures should also be taken for the buildings located on the medium soft site soil.

3.2.2   Calculation of seismic bearing capacity of foundation soil

In order to determine the seismic bearing capacity of foundation soil, it is necessary to study the strength of soil under dynamic load, that is, the dynamic strength of soil (abbreviated as dynamic strength). Generally, under the action of dynamic load and static load, dynamic strength refers to the total stress when the soil specimen reaches a certain strain value (the limit strain value of the static load) at a certain number of cycles of dynamic load. Therefore, it is related to static load, pulse number, frequency, allowable strain and so on. Because earthquakes are limited ($10\sim30$) pulses of low frequency ($1\sim5\text{Hz}$), the dynamic strength of most soils is higher than that of static strength, except for weak soil. In addition, considering that the earthquake is a kind of accidental action and the duration is short, the reliability requirement of the foundation under seismic action can be lower than that under static action. In this way, the seismic bearing capacity of foundation soil is higher than the static bearing capacity except very weak soil. In China and in most countries in the world, the value of seismic bearing capacity of foundation soil is determined by multiplying the static bearing capacity of foundation soil by an adjustment coefficient. The seismic bearing capacity of base shall be calculated according to the following formula:

$$f_{aE} = \zeta_a f_a \tag{3-4}$$

Where f_{aE} ——adjusted seismic bearing capacity of base;

ζ_a ——adjustment coefficient of base seismic bearing capacity, which shall be adopted according to Table 3.7;

f_a ——characteristic value of base bearing capacity after depth and width correction, which shall be adopted according to the current national standard 《Code for Design of Building Foundation》.

The seismic bearing capacity of foundation soil is generally higher than that of static bearing capacity of foundation soil, which can be explained from the point of view that elastic deformation of foundation soil is only considered but permanent deformation is not considered under earthquake action.

Adjustment coefficient of the seismic bearing capacity of base   Table 3.7

Name and character of rock-soil	ζ_a
Rock, dense detritus, dense gravel, course and medium sands, cohesive soil and silt with $f_{ak} \geqslant 300$kPa	1.5
Medium dense and slightly dense detritus, medium dense and slightly dense gravel, course and medium sands, dense and medium dense fine and mealy sands, cohesive soil and silt with 150kPa$\leqslant f_{ak} < 300$kPa, and hard loess	1.3
Slightly dense fine and mealy sands, cohesive soil and silt 100kPa$\leqslant f_{ak} < 150$kPa, and plastic loess	1.1
Mud, muddy soil, loose sand, miscellaneous fill soil, newly piled loess and streamed loess	1.0

In order to check the vertical bearing capacity of natural foundation under earthquake action, the pressure of foundation ground is taken as a straight line distribution (Figure 3.1). According to the standard combination of seismic effect, the average pressure at the bottom and the maximum pressure at the edge of foundation should meet the requirements of the following formulas:

$$p \leqslant f_{aE} \tag{3-5}$$

$$P_{max} \leqslant 1.2 f_{aE} \tag{3-6}$$

Where p ——mean pressure on foundational bottom according to the standard combination of earthquake action effects;

P_{max} ——maximum pressure on foundation edge according to the standard combination of earthquake action effects.

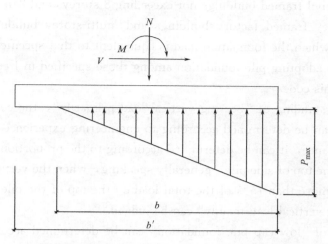

Figure 3.1   Pressure distribution on base bottom

The *Code for Seismic Design of Buildings* GB 50011—2010 stipulates, for the tall buildings with height-width ratio above 4, the foundation bottom should not appear with any abscission zone (zero stress zone) under the earthquake action; for the other buildings, the area of the abscission zone (zero stress zone) between the foundation bottom and foundation soil shall not exceed 15% of the area of foundation bottom. According to the latter rule, the ratio of compression width to foundation width should be more than 85% for foundation with rectangular bottom. That is:

$$b' \geqslant 0.85 \, b \tag{3-7}$$

Where b' ——the compression width of rectangular base (Figure 3.1);

b ——width of bottom of rectangle base.

3.2.3   Pile foundations

In the seismic design of pile foundation, how to realize the fortification target of "strong earthquake will not collapse" has always been concerned by engineering circles. Before the relevant codes are clearly stipulated, proper leeway should be left for the design of foundation piles for major and special projects, and effective structural measures should be taken to strengthen the connection between piles and caps so as to ensure that the connections do not fail in the event of a strong earthquake.

The following buildings, of which the low-cap pile foundation mainly bearing vertical loads and there is no liquefied soil layer under the ground, no mud or muddy soil and no filled soil with characteristic value of base bearing capacity no larger than 100 kPa surrounding the pile cap, may not be carried out with the seismic bearing capacity check of pile foundation:

(1) The following buildings for Intensity 7 and 8:

1) Ordinary single-storey factory buildings and single-storey spacious buildings;

2) Ordinary civil framed buildings not exceeding 8 storeys and 24m in height;

3) Multi-storey framed factory buildings and multi-storey buildings with concrete seismic wall, for which the foundation load is equivalent to that specified in Item 2).

(2) Buildings adopting pile foundation among those specified in Item 1 and Item 3 of Article 4. 2. 1 of this code.

For the understanding of "bearing vertical load primarily", the code has no specific provisions and it can be determined according to engineering experience. When there is no engineering experience, it can be determined according to the proportion of vertical load to the total load on the top of side pile, generally speaking, when the vertical load on the top of the side pile is more than 75% of the total load on the top of the pile, the pile foundation with "bearing vertical load mainly" can be judged.

The criterion of "low cap pile foundation" can be determined according to the engineering experience, and when there is no engineering experience, it can be determined according to the contact relationship between the pile bottom and the foundation soil. When the bottom of the cap is in close contact with the natural foundation (the settlement of the foundation should be considered), it can be determined as "low cap pile foundation" (the "cantilever pile" will not be formed when the pile actually works).

3.3 Liquefied soil base

3.3.1 Concept of liquefaction

The saturated loose sand and silt below the groundwater level tend to become denser under seismic action, but the soil particles are suspended because the pore water can not be discharged, forming a liquid-like state. This phenomenon is called soil liquefaction (Figure 3.2).

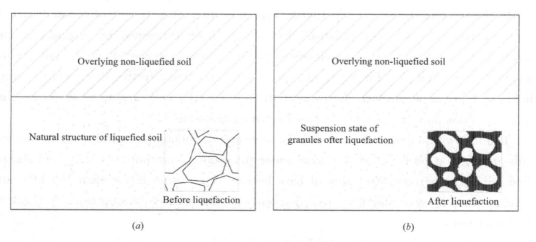

(a)　　　　　　　　　　　　　　　　(b)

Figure 3.2   Schematic diagram of liquefaction of soil

The Xingtai earthquake in 1966, the Haicheng earthquake in 1975 and the Tangshan earthquake in 1976 in China all had liquefaction phenomena, which caused the buildings to be damaged to varying intensities. According to the principle of soil mechanics, the liquefaction of sand is due to the zero shear strength of saturated sand in a short period of time during earthquake. We know that the shear strength of saturated sand can be written as:

$$\tau_f = \bar{\delta} \tan\phi = (\delta - u) \tan\phi \tag{3-8}$$

where $\bar{\delta}$ ——effective normal compressive stress (intergranular compressive stress) on shear plane;

δ ——total normal compressive stress on shear plane;

u ——pore water pressure on shear plane;

ϕ ——internal friction angle of soil.

3.3.2 Evaluation of liquefaction soil

The liquefaction evaluation of saturated sandy soil and silt (excluding loess) and the base treatment may not be carried out under conditions for Intensity 6, however, those of the Category B buildings that are sensitive to liquefaction settlement shall be carried out according to requirements for Intensity 7 and those of Category B buildings for Intensity 7~9 may be carried out according to the requirements of local seismic precautionary intensity.

If saturated sandy soil and saturated silt exist under ground, buildings except for those for Intensity 6, all shall be carried out with the liquification evaluation; for the base with liquefied soil layer, corresponding measures shall be taken according to the precautionary category of building and liquification degree of base, in combination of the specific conditions.

The saturated sandy soil or silt (excluding loess), if meeting any one of the following conditions, may be preliminarily evaluated as non-liquefaction or the influence of liquification may not be considered.

(1) The geologic time of soil is Epipleistocene of Quaternary (Q3) or earlier, and they may be evaluated as non-liquefied soils for Intensity 7 and 8.

(2) If the percentage content of sticky particles (particles with grain size less than 0.005mm) in silt is not less than 10%, 13% and 16% for Intensity 7, 8 and 9 respectively, the soil may be evaluated as non-liquefied soil.

Note: the sticky particle content used for liquification evaluation is measured by using hexametaphosphate as dispersion agent, and it shall be converted according to relevant regulations if other methods are adopted.

(3) For buildings with shallowly-buried natural base, if the thickness of upper covered non-liquefied soil and the depth of underground water level meet any one of the following conditions, the liquification influence may not be considered.

$$d_u > d_0 + d_b - 2 \tag{3-9}$$

$$d_w > d_0 + d_b - 3 \tag{3-10}$$

$$d_u + d_w > 1.5d_0 + 2d_b - 4.5 \tag{3-11}$$

Where d_w ——depth of underground water level (m), which should be adopted according to the mean annual highest water level within the design reference period or may be adopted according to the annual highest water level in recent years.

d_u ——thickness of upper covered non-liquefied soil layer (m), in which the thickness of mud and muddy soil layers should be deducted.

d_b ——embedded depth of foundation (m), which shall be 2m if it is less than 2m.

d_0 ——characteristic depth of liquefied soil (m), which may be adopted according to Table 3.8.

Characteristic depth of liquefied soil (m)　　　　**Table 3.8**

Type of saturated soil	Intensity 7	Intensity 8	Intensity 9
Silt	6	7	8
Sandy soil	7	8	9

Note: If the underground water level in this regions is under variable condition, the characteristic depth shall be considered according to unfavorable conditions.

If the preliminary discrimination of saturated sandy soil and silt indicates further liquification evaluation is necessary, standard penetration test shall be adopted to discriminate the liquification condition of the soil within 20m (deep) under the ground; but for the buildings that may not be carried out with seismic capacity check of natural base and foundation as specified in Article 3.3.2, the liquification condition of the soil within only 15m (deep) under the ground may be evaluated. If the standard penetration blow count of saturated soil (without pole length correction) is less than or equal to the standard penetration blow count for liquification evaluation, the soil shall be evaluated as liquefied soil. If mature experiences are available, other evaluation methods may also be adopted.

Within the depth range of 20m under the ground, the critical value of standard penetration blow count for liquification evaluation may be calculated according to following formula:

$$N_{cr} = N_0 \beta \left[\ln(0.6d_s + 1.5) - 0.1d_w\right] \sqrt{\frac{3}{\rho_c}} \tag{3-12}$$

Where N_{cr} ——critical value of standard penetration blow count for liquification evaluation;

N_0 ——reference value of standard penetration blow count for liquification evaluation, which may be adopted according to Table 3.9;

d_s ——depth of standard penetration point for saturated soil (m);

d_w ——underground water level (m);

ρ_c ——percentage content of sticky particles, which shall be taken as 3 if it is less than 3 or the soil is sandy soil;

β ——adjustment coefficient, which shall be taken as 0.80 for design earthquake Group 1, 0.95 for Group 2 and 1.05 for Group 3.

Reference value of standard penetration blow count for liquification evaluation　Table 3.9

Design basic acceleration of ground motion(g)	0.10	0.15	0.20	0.30	0.40
Reference value of standard penetration blow count for liquification evaluation	7	10	12	16	19

【**Example 3.3**】

A foundation profile map of a site is shown in Figure 3.3. Thickness of non-lique-fied soil layer $d_u = 5.5$m. The bottom is sand. Depth of under ground water level $d_w = 6.0$m. Embedded depth of foundation $d_b = 2$m. The site is intensity 8. According to the pre-liminary discriminant formula and Figure 3.3, whether the effect of liquefaction is consid-ered.

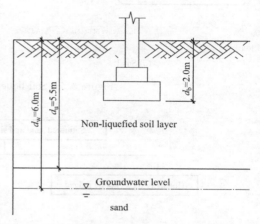

Figure 3.3　Appended drawings

Solution:

(1) According to the discriminant formu-la (3-11), the characteristic depth of liquefied soil was found in Table 3.8. So $d_0 = 8$m, Be-cause

$$1.5 d_0 + 2 d_b - 4.5 = 1.5 \times 8 + 2 \times 2 - 4.5 = 11.5\text{m} = d_u + d_w = 5.5 + 6 = 11.5\text{m}$$

The calculation shows that the sides of the formula (3-11) are exactly equal on both sides. Therefore, this example will further distinguish the liquefaction effect of sand soil layer.

(2) Find out on the abscissa axis of Figure 3.4, $d_u = 5.5$m, Find out on ordinate ax-is, $d_w = 6$m, their intersection points are located at the Intensity 8 diagonal line respec-tively. It is indicated that the influence of liquefaction of the sand layer should be further judged.

3.3.3　Evaluation of liquefied foundation

Flow chart for liquefaction evaluation is shown in Figure 3.5.

1. Significance of evaluation

In the past, two conclusions of liquefaction or non liquefaction were given according to the discriminant formula. Therefore, the liquefaction hazard can not be quantitatively e-valuated, and the corresponding anti-liquefaction measures can not be taken. The degree of

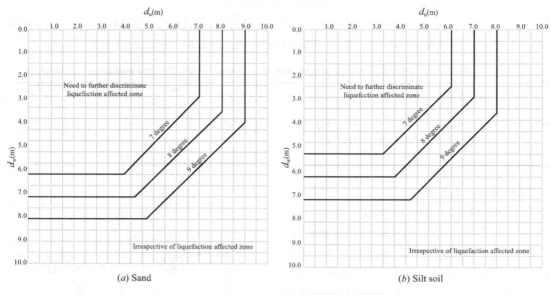

Figure 3. 4　Soil liquefaction discrimination map

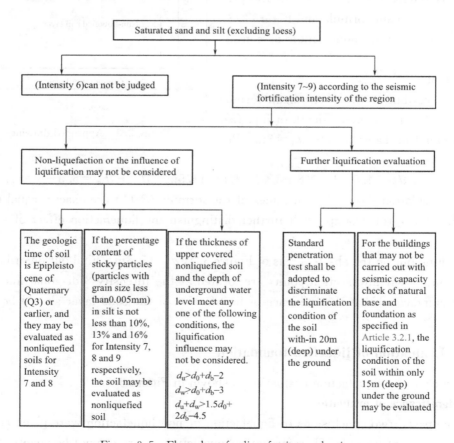

Figure 3. 5　Flow chart for liquefaction evaluation

liquefaction on the foundation is different，and the harm to the building is also different. Therefore，the analysis and evaluation of the hazards of liquefied foundation is a very important issue in the seismic design of buildings.

2. Liquification index

For the base with liquefied sandy soil layer and silt layer，the depth and thickness of the liquefied soil layers shall be explored，the liquification index of each drilling hole shall be calculated according to the following formula，and the liquification grade of base shall be comprehensively classified according to Table 3. 10.

$$I_{IE} = \sum_{i=1}^{n} \left[1 - \frac{N_i}{N_{cri}} \right] d_i W_i \qquad (3-13)$$

Where　I_{IE}──liquification index；

n──total number of standard penetration test points for each drilling hole within the evaluated depth range；

N_i、N_{cri}──measured value and the critical value of the standard penetration blow count at the ith point respectively，and the critical value shall be taken if the measured value is larger than the critical value；if the liquification within a scope of 15m only needs to be evaluated，the measured value bellow 15m may be adopted according to critical value；

d_i──thickness of soil layer represented by the ith point，which may be taken as half the difference between the depth values at these two upper and lower standard penetration test points adjoining this standard penetration test point，however，its upper limit shall not be larger than the depth of underground water level and its lower limit shall not be deeper than the liquification depth；

w_i──weight horizon influence function value（unit：m^{-1}）of unit thickness of the ith soil layer，which shall be taken as 10 if the depth of midpoint of this layer is no larger than 5m，zero if the depth is equal to 20m，and be taken with value with linear interpolation if the depth is from 5~20m.

Corresponding relationship between liquification grade and liquification index Table 3. 10

Liquification grade	Light	Moderate	Serious
Liquification index	$0 < I_{IE} \leqslant 6$	$6 < I_{IE} \leqslant 18$	$I_{IE} > 18$

【**Example 3. 4**】According to Intensity 8 fortification，the engineering geological age of a certain project belongs to the following：sand layer to 2. 1m，gravel sand layer to 4. 4m，fine sand to 8. 0m，silty clay layer to 17m；clay content of sand layer and fine sand layer is lower than 3%；groundwater level depth is 1. 0m；foundation buried depth is 1. 5m；design seismic site group belongs to the first group. The basic acceleration is 0. 2g. The results of the test are shown in Table 3. 11. and may be evaluated for liquefaction of the site.

Solution:

(1) Initial judgment

$$Q_4 : d_0 + d_b - 3 = 7 > 1 = d_w$$
$$d_u = 0$$
$$1.5d_0 + 2d_b - 4.5 = 11.5 \geqslant d_w + d_u$$
$$\rho_c < 13$$

Therefore, the non-liquefaction conditions are not satisfied and need to be further e-valuationed.

(2) Standard penetration test evaluation

1) N_{cr} is calculated according to formula (3-12), in which $N_0 = 12$ and $d_w = 1.0$. The thickness of the layer represented by each standard point of penetration is given in the question. The calculation results are shown in Table 3.11, and point 4 are nonliquefied soil.

2) Calculate layer displacement impact function:

The first point, the groundwater level is 1.0m, so the upper limit is 1.0m and the soil thickness is 1.1m. So

$$Z_1 = 1.0 + \frac{1.1}{2} = 1.55 \text{ , } w_1 = 10$$

The second point, the upper boundary is gravel layer, the deep bottom is 4.4m, representing the thickness of the soil layer is 1.1m. So

$$Z_2 = 4.4 + \frac{1.1}{2} = 4.95 \text{ , } w_1 = 10$$

3) The liquefaction index of each layer is calculated according to formula (3-13), and the results are shown in Table 3.11.

Finally $I_{lE} = 10.95$, according to Table 3.10, the liquefaction grade is medium.

<table>
<tr><td colspan="8">Liquefaction analysis table</td><td>Table 3.11</td></tr>
</table>

Measuring point	Point depth d_{si} (m)	Standard intersecting value N_i	Thickness of measured soil layer d_i (m)	Critical value of standard penetration N_{cri}	d_i Midpoint depth Z_i (m)	W_i	I_{lE}
1	1.4	5	1.1	7.2	1.5	10	3.36
2	5.0	7	1.1	13.4	4.95	10	5.25
3	6.0	11	1.0	14.7	6.0	9.3	2.34
4	7.0	16	1.0	15.7			

3.3.4 Anti-liquefaction measures

The anti-liquefaction measures of foundation should be determined comprehensively according to the types of anti-seismic fortification of buildings, the liquefaction grade of foundation and the concrete conditions. When the liquefied sand and silt layers are flat and uniform, the anti-liquefaction measures of foundation should be selected according to Table 3. 12. The influence of gravity load of superstructure on liquefaction hazard can also be considered，and the anti-liquefaction measures should be adjusted according to the estimation of liquefaction seismic subsidence. The untreated liquefied soil should not be used as the bearing layer of natural foundation.

Anti-liquefaction measures | Table 3. 12

Seismic precaution of building	Liquification grade of base		
Precautionary category	Light	Moderate	Serious
B	Eliminating the liquefaction settlement partially or treating the foundation and superstructure	Eliminating the liquefaction settlement wholly or eliminating the liquefaction settlement partially and treating the foundation and superstructure	Eliminating the liquefaction settlement wholly
C	Treating the foundation and superstructure, or taking no measures	Treating the foundation and superstructure, or taking measures of much higher requirement	Eliminating the liquefaction settlement wholly, or eliminating the liquefaction settlement partially and treating the foundation and superstructure
D	May not take measures	May not take measures	Treating the foundation and superstructure. or taking other economical measures

The specific requirements for anti liquefaction measures in the Table 3. 12 are as follows：

1. The measures for eliminating the liquefaction settlement of base wholly shall meet the following requirements：

(1) When pile foundation is used，the length (pile-tip not included) of the pile tip driven into the stable soil layer below the liquefaction depth shall be determined through calculation，which shall not be less than 0. 8m for detritus，gravel，coarse and medium sands，stiff cohesive soil，and dense silt and should not be less than 1. 5m for other non-rocky soil.

(2) When deep foundation is used，the bottom of foundation shall be embedded in the stable soil layer below the liquefaction depth，and the embedded depth shall not be less than 0. 5m.

(3) When a compaction method (e. g. vibroflotation，vibration compaction，gravel pile compaction，and dynamic compaction) is used for strengthening，compaction shall be carried out down to the lower margin of liquefaction depth；after strengthening the gravel piles by vibroflotation or compaction，the standard penetration blow count of soil between piles should not be less than the critical value of standard penetration blow count for liquefication evaluation as specified in Article 4. 3. 4 of the *Code for Seismic Design of Buildings*

GB 50011—2010.

(4) The totally liquefied soil layer shall be replaced with non-liquefied soil, or the thickness of the upper covered non-liquefied soil layer shall be increased.

(5) When treating by adopting compaction method or soil replacement method, the treatment width outside the foundation edge shall exceed 1/2 of the treatment depth bellow the foundation bottom and shall not be less than 1/5 of the foundation width.

2. The measures for partially eliminating the liquefaction settlement of base shall meet the following requirements:

(1) The treatment shall be carried out to a depth so that the liquefaction index of base is reduced after treatment, and the liquefaction index should not be larger than 5; for central zone of large area raft foundation and box foundation, the liquefaction index after treatment hereof may be reduced to 4; for individual foundation and strip foundation, the liquification index also shall not be less than the larger value between characteristic depth of liquefied soil under the foundation bottom and the foundation width.

Note: The central zone refers to the zone located within the foundation outer edge, and within over 1/4 length of the corresponding direction along the length/width direction away from the outer edge.

(2) After strengthening through vibroflotation or gravel pile compaction, the standard penetration blow count of soil between piles should not be less than the critical value of standard penetration blow count for liquification evaluation as specified in Article 4.3.4 of the *Code for Seismic Design of Buildings* GB 50011—2010.

(3) The treatment width outside the foundation edge shall meet the requirements of Clause 5 in Article 4.3.7 of the *Code for Seismic Design of Buildings* GB 50011—2010.

(4) Other measures to reduce seismic subsidence due to liquefaction shall be adopted, like thickening the upper covered nonliquefied soil layer and improving the peripheral drainage condition, etc..

3. The treatment of foundation and superstructure to reduce the liquification affect may be adopted with the following measures comprehensively:

(1) Appropriate embedded depth of foundation shall be selected.

(2) The bottom area of foundation shall be adjusted to reduce the eccentricity of foundation.

(3) The integrity and stiffness of foundation shall be strengthened, for instance, a-dopting box foundation, raft foundation or reinforced concrete cross strip foundation, installing foundation ring beams in addition, etc..

(4) The loads shall be reduced, the integral stiffness and uniform symmetry of superstructure shall be reinforced, the settlement joints shall be arranged reasonably and the adoption of such structure form sensitive to differential settlement shall be avoided.

(5) Adequate size shall be reserved at the position where pipelines pass through the buildings, otherwise the flexible joints shall be adopted.

Exercises

3. 1 What is the equivalent shear wave velocity of the soil layer? How to calculate?

3. 2 What is the site cover thickness? How to determine?

3. 3 What is the excellence cycle of the site?

3. 4 What is the liquefaction of foundation soil? What are the hazards?

3. 5 How to determine the liquefaction of foundation soil?

3. 6 The construction site drilling known geological data are shown in Table 3. 13，determine the site category.

Borehole data   Table 3. 13

Bottom depth of upper layer(m)	Thickness of soil layer(m)	Name of rock and soil	Shear wave velocity of soil layer(m/s)
2. 00	2. 00	Miscellaneous filled soil	210
5. 50	3. 50	Silty soil	280
9. 50	4. 00	Medium sand	360
15. 50	6. 00	Gravel sand	540

Chapter 4　Dynamics of Structures and Seismic Response

4. 1　Introduction

Most loads that occur on a structure can be considered as static (time independent) or quasi-static (time dependent, but slow enough such that inelastic effects can be ignored) loads, such as dead loads or live loads on roofs, which require only static analysis. Although all loads other than dead loads are transient, it is customary in most designs to treat these loads as static.

Seismic design is a dynamic (time dependent) problem. The term " dynamic" simply refers to " time varying". A dynamic load is one, the magnitude, direction, or point of application of which varies with time. The structural response to a dynamic load, i. e. , the resulting deflections or stresses, is also time dependent to dynamic. In general, the structural response to any dynamic loading is expressed in terms of the displacements of the structure.

In terms of confidence in their values, dynamic loads may be classified into deterministic (or prescribed) and stochastic (or random). If the loading is a known function of time, the loading is said to be prescribed, and the analysis of a structural system to a prescribed loading is called deterministic analysis. In contrast, the variations of a random force in time may be affected by a number of factors, so its determination always implies a certain probabilistic element. Seismic loads are random in character, though they are usually regarded as deterministic in practical calculations to simplify the design model.

Dynamic loading may also be classified as periodic loads [Figure 4. 1 (a) and (b)] or non-periodic loads [Figure 4. 1 (c) and (d)]. Periodic loads are examples of repetitive loads exhibiting time variation successively for a large number of cycles. The simplest periodic load is the sinusoidal or the cosine variation [Figure 4. 1 (a)], termed as simple harmonic. Non-periodic loads may be either short-duration impulsive loads, as shown in Figure 4. 1 (c), or long-duration loads [Figure 4. 1 (d)]. Blasts and explosions are examples of short-duration impulsive loads. Earthquakes and wind are examples of long-duration loads.

Dynamics deals with the motion of nominally rigid bodies. Structural dynamics implies that in addition to having motion, the bodies are non-rigid. In a structural-dynamic problem, the load and response vary with time; hence, a dynamic problem does not have a

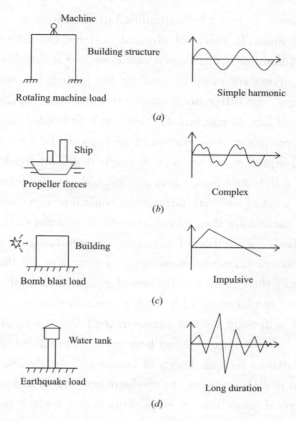

Figure 4. 1   Typical dynamic loads

single solution. Since earthquake forces are considered dynamic, instead of obtaining a single solution as in a static case, a separate solution is required at each instant of time for the entire duration of the earthquake. When a dynamic load p (t) is applied to a structure, e. g., on a simple beam as shown in Figure 4. 2, the resulting displacements are associated with accelerations that produce inertia forces resisting the accelerations. Thus the internal moments and shears in the example structure (beam) of Figure 4. 2 must equilibrate not only the externally applied force but also the inertia forces resulting from the acceleration of the beam. These inertia forces cause the system to vibrate. Structural dynamics, however, should not be confused with vibration, which implies only oscillatory behaviour.

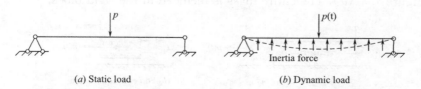

(*a*) Static load   (*b*) Dynamic load

Figure 4. 2   Basic difference between static and dynamic loads

Behaviour of a system subjected to dynamic load is quite complex. However, it can be determined with sufficient accuracy by a simplified mathematical model of the system that may be linear or non-linear. In fact, all physical systems exhibit non-linearity. Accurate modelling will lead to non-linear differential equations and it is difficult to solve and find their solution. Assumptions are made to linearise the system and with application of the principles of dynamics, the differential equations governing the behaviour can be obtained. Newton's second law of motion, D'Alembert's principle, principle of virtual displacement, and the principle of conservation of energy may be used suitably to derive the governing differential equations of motion. A single-degree-of-freedom (SDOF) system leads to one ordinary differential equation of motion and a multi-degrees-of-freedom (MDOF) system leads to a set of ordinary differential equations of motion. The governing differential equations of motion are then solved to find the response of the system. Techniques such as classical methods for solution of differential equations, time-domain method, frequency-domain method, or numerical methods as appropriate for the particular case may be used. The solution of the governing differential equations of motion gives the displacements, velocities, and accelerations of various masses in dynamic analyses. The ultimate aim of the analysis is to develop a set of curves in the form of response spectrum.

In the following sections, attempt has been made to bring out the essentials of structural dynamics as related to seismic design of buildings, the dynamic analysis consists of defining the analytical model, deriving the mathematical model and solving for the dynamic response. Mathematical modelling of single-storey and multi-storeys structures, with and without damping, is presented briefly.

4. 2　Seismic response analysis of single-degree-of-freedom systems

4. 2. 1　Systems with single degree of freedom

The physical properties of any linearly elastic structural system subjected to dynamic loads include its mass m measured in kg or ton (should not be confused with its wight $m \times g$, which is a force measured in N or kN), its elastic properties (flexibility or stiffness), stiffness k, measured in N/m or kN/m, its energy-loss mechanism (damping) denoted by c, the constant of proportionality between force and velocity measured in N $\cdot$ s/m or kN $\cdot$ s/m, and the external source of excitation (load). A sketch of such a system is shown in Figure 4. 3 (a). The entire mass is included in the rigid block.

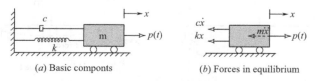

(a) Basic componts　　　　　(b) Forces in equilibrium

Figure 4. 3　Idealized SDOF system

Rollers constrain this block so that it can move only in simple translation; thus the single displacement coordinate v completely defines its position. The elastic resistance to displacement is provided by the weightless spring of stiffness k, while the energy-loss mechanism is represented by the viscous damper (also known as dashpot) c. The external-loading mechanism producing the dynamic response of this system is the time-varying load p (t).

As shown in Figure 4. 3 (b), for the applied load $p(t)$ the resulting forces are inertia force f_I, damping force f_D, and the elastic spring force f_S. The equation of motion for this system is given as:

$$f_I + f_D + f_S + p(t) = 0 \tag{4-1}$$

Where　f_S (elastic force) = spring stiffness×displacement= $-kx$ (for a linear system);

f_I (inertia force) =mass×acceleration= $-m\ddot{x}$;

f_D (damping force) = damping constant×velocity= $-c\dot{x}$;

$p(t)$ =the force p varying with time.

The negative sign indicates that the direction of force is opposite to that of the x-axis. Equation (4-1) can be written as:

$$m\ddot{x} + c\dot{x} + kx = p(t) \tag{4-2}$$

The equation of motion developed above is for a displacement x of the idealized structure of Figure 4. 3 (a), assumed to be linearly elastic and subjected to an external dynamic force $p(t)$. In the inelastic range, the force f_S corresponding to deformation x depends on the history of the deformation and on whether the deformation is increasing (positive velocity) or decreasing (negative velocity). Thus the resisting force can be expressed as $f_S(x, \dot{x})$. The derivation of equation of motion for elastic systems can be extended to inelastic systems where the equation of motion becomes

$$m\ddot{x} + c\dot{x} - f_S(x, \dot{x}) = p(t) \tag{4-3}$$

In this book, only the problems in the elastic range are researched.

4. 2. 2　Dynamic response of single-storey structure

A single-storey structure can be modeled as a SDOF system. Each possible displacement of the structure is known as degree of freedom. For linear dynamic analysis, a structure can be defined by the three key properties——the mass, the stiffness, and the damping. The mass m of the structure is assumed to be concentrated at the floor level of the storey. The horizontal girder in the frame is assumed to be rigid and to include all the moving mass of the structure as shown in Figure 4. 4. In reality, all structures have distributed mass, stiffness, and damping. Assumption of lumped mass, however, is justified as in most cases, it is possible to obtain reasonably accurate estimates of dynamic behaviour of the structure. The vertical columns are assumed to be weightless and inextensible in the vertical (axial) direction. The resistance to girder displacement provided by each column is represented by its spring constant $k/2$ (since there are two columns). The idealized structure has only one DOF, the lateral displacement x, since it has been idealized with mass concentrated at one location (roof level) for dynamic analysis, x is associated with column flexure; the damper c (represented by dashpot) provides a velocity-proportional resistance to this deformation. The system is called SDOF system. Then, the equation of motion for a single-storey structure will be

$$f_I + f_D + f_S = 0 \qquad (4\text{-}4)$$

or

$$m\ddot{x} + c\dot{x} + kx = 0 \qquad (4\text{-}5)$$

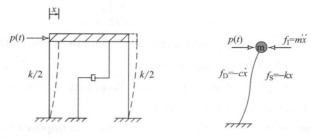

Figure 4.4 An SDOF system under horizontal force

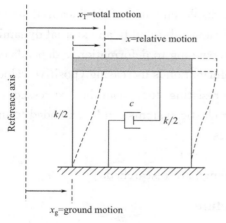

Figure 4.5 An SDOF system under ground motion

The dynamic stresses and deflections may be induced in a structure not only by a time-varying applied load but also by motions of its support points and the motions of the building's foundation caused by an earthquake. Figure 4.5 shows a simplified model of the earthquake excitation problem in which the horizontal ground motion displacement caused by the earthquake is indicated by the displacement x_g of the structure base relative to the fixed reference axis. In this case, the structure is subjected to ground acceleration, and total displacement of the mass at any instant can be expressed as the sum of the ground displacement x_g and the column distortion x.

$$f_I = -m(\ddot{x} + \ddot{x}_g) \qquad (4\text{-}6)$$

$$f_D = -c\dot{x} \qquad (4\text{-}7)$$

The damping coefficient c may be determined experimentally by conducting vibration experiments on actual structure, which is nonrealistic.

Assuming the relationship of force f_S and deformation x to be linear (i. e. , loading and unloading curves to be identical), the system has to be elastic. Hence for linearly elastic system

$$f_S = -kx \qquad (4\text{-}8)$$

From Equations (4-4) to (4-8)

$$m(\ddot{x} + \ddot{x}_g) + c\dot{x} + kx = 0 \qquad (4-9)$$

or

$$m\ddot{x} + c\dot{x} + kx = -m\ddot{x}_g \qquad (4-10)$$

The equation of motion developed above governs the relative displacement x of the structure of Figure 4.5 subjected to ground acceleration $\ddot{x}_g$. The negative sign in Equation (4-10) indicates that the effective force opposes the direction of ground acceleration; in practice this has little significance, in as much as base input must be assumed to act in an arbitrary direction.

A comparison of Equations (4-2) and (4-10) shows that the equations of motion for the structure are subjected to two separate excitations at each instant of time——ground acceleration $\ddot{x}_g$ and external force $-m\ddot{x}_g$ are one and the same. Thus the relative displacement of deformation x of the structure due to ground acceleration $\ddot{x}_g$ will be identical to the displacement x of the structure if its base were stationary and if it were subjected to an external force equal to $-m\ddot{x}_g$. The ground motion can, therefore, be replaced by the effective earthquake force as shown in Figure 4.6.

$$P_{eff} = -m\ddot{x}_g \qquad (4-11)$$

This force is equal to mass times the ground acceleration, acting opposite to the acceleration.

It is important to recognize that the effective earthquake force is proportional to the mass of the structure. Thus the effective earthquake force increaseswhen the structural mass increases

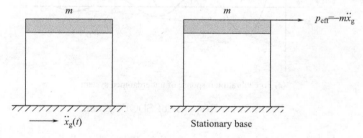

Figure 4.6　Effective earthquake force

1. Free vibration response

Motions taking place with the applied force set equal to zero are called free vibrations. To establish the free vibration response of the system, let us assume, firstly, that there is no ground motion and that the SDOF system is without damping. Under these conditions, the system is in motion and is governed only by the influence of the so called initial conditions; that is, the given displacement $x(0)$ and velocity $\dot{x}(0)$ at time $t=0$ when the study of the system is initiated. Equation (4-10) can be simplified to

$$m\ddot{x} + kx = 0 \qquad (4-12)$$

Subject to these initial conditions, the solution to the homogeneous differential equation is obtained by standard methods

$$x(t) = x(0)\cos\omega t + \frac{\dot{x}(0)}{\omega}\sin\omega t \qquad (4-13)$$

where $\omega = \sqrt{\dfrac{k}{m}}$ ———circular frequency or angular velocity of the system.

This solution represents a simple harmonic motion and is shown in Figure 4. 7 (a).

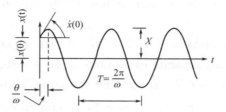

(a) Undamped free vibration response

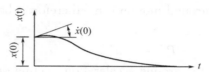

(b) Free vibration response with critical damping

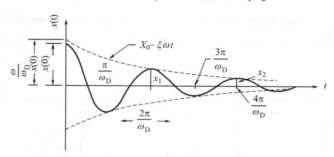

(c) Free vibration response of underdamped system

Figure 4. 7   Response of SDOF system

The natural period T defined as the time required for the phase angle ωt to travel from 0 to 2π is given by

$$T = \frac{2\pi}{\omega} = 2\pi\sqrt{\frac{m}{k}} \qquad (4-14)$$

Equation (4-13) can be rewritten as follows

$$x(t) = X\cos(\omega t + \theta) \qquad (4-15)$$

where

$$X = \sqrt{[x(0)]^2 + \left[\frac{\dot{x}(0)}{\omega}\right]^2} \qquad \text{(amplitude)} \qquad (4-16)$$

$$\theta = \arctan\left[\frac{\dot{x}(0)}{\omega x(0)}\right] \qquad \text{(phase angle)} \qquad (4-17)$$

This is shown in Figure 4.8 the phase angle represents the angular distance by which the resultant motion lags behind the cosine term in the response.

In real systems, energy may be lost as a result of damping. Then, the free vibration response of a damped SDOF system will diminish with time. If viscous damping is present, the equation of motion will be

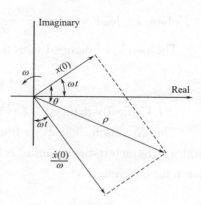

$$m\ddot{x} + c\dot{x} + kx = 0 \tag{4-18}$$

Dividing it by m, we can get

$$\ddot{x} + 2\xi\omega\dot{x} + \omega^2 x = 0 \tag{4-19}$$

Figure 4.8   Rotating-vector representation of free vibrations

Where $\xi = \dfrac{c}{2m\omega}$, $\omega = \sqrt{\dfrac{k}{m}}$, and ξ is the damping factor also known as damping ratio or fraction of critical damping.

Substituting $x = A e^{\lambda t}$, $\dot{x} = A\lambda e^{\lambda t}$, and $\ddot{x} = A\lambda^2 e^{\lambda t}$ in Equation (4-19), where "e" is natural constants (Euler number).

$$A e^{\lambda t}(\lambda^2 + 2\xi\omega\lambda + \omega^2) = 0 \tag{4-20}$$

The roots of above characteristic equation are

$$\lambda_1, \lambda_2 = \omega(-\xi \pm \sqrt{\xi^2 - 1}) \tag{4-21}$$

The solution of Equation (4-19) is

$$x = A e^{\lambda_1 t} + B e^{\lambda_2 t} \tag{4-22}$$

Equation (4-19) indicates that the solution changes its form according to the magnitude of the damping factor ξ. there can be three cases:

(1) Under damped ($0 < \xi < 1$) [Figure 4.7 (c)]

The motion is oscillatory with decaying amplitude [Figure 4.7 (c)]. For this case in which damping is less than the critical, the solution of equation is of the form

$$x = \exp(-\xi\omega t)[A \cos\omega_\mathrm{D} t + B \sin\omega_\mathrm{D} t] \tag{4-23}$$

Where A and B are constants of integration; and $\omega_\mathrm{D} = \omega\sqrt{1 - \xi^2}$ is the damped frequency of the system. The value of damping ratio of real structures is very small, and usually ranges between 2%~20%. For example, the damping ratio of reinforced concrete structures is 0.05.

A and B can be evaluated from the initial conditions of displacement $x(0)$ and velocity $\dot{x}(0)$, and substituted into Equation (4-23) gives the solution of damped free vibration and can be expressed as

$$x = \exp(-\xi\omega t)\left[x(0)\cos\omega_\mathrm{D} t + \frac{\dot{x}(0) + x(0)\xi\omega}{\omega_\mathrm{D}}\sin\omega_\mathrm{D} t\right] \tag{4-24}$$

Alternatively, this expression can be written as

$$x = \rho\exp(-\xi\omega t)\cos(\omega_\mathrm{D} t + \theta) \tag{4-25}$$

Where amplitude: $\rho = \sqrt{[x(0)]^2 + \left[\dfrac{\dot{x}(0) + x(0)\xi\omega}{\omega_\mathrm{D}}\right]^2}$ \hfill (4-26) phase angle: $\theta = \arctan\left[\dfrac{\dot{x}(0) + x(0)\xi\omega}{x(0)\omega_D}\right]$ (4-27)

The period of damped vibration is given by

$$T_D = \frac{2\pi}{\omega\sqrt{1-\xi^2}} = \frac{T}{\sqrt{1-\xi^2}} \tag{4-28}$$

(2) Critically damped ($\xi = 1$) [Figure 4.7 (b)]

This condition indicates a limiting value of damping at which the system loses its vibratory characteristics; this is called critical damping. The damping coefficient at critical damping is denoted by

$$c_c = 2\sqrt{mk} \tag{4-29}$$

In a critically damped system, the roots of the characteristic equation are equal. The general solution for a critically damped system would be

$$x = \{x(0) + [\dot{x}(0) + x(0)\sqrt{\frac{k}{m}}]t\}\exp(-\sqrt{\frac{k}{m}}t) \tag{4-30}$$

(3) Over damped ($\xi > 1$)

For this case, the response is similar to the motion of the critically damped system of Figure 4.7 (b), but the return towards the neutral position requires more time as the damping ratio is increased. The system does not oscillate because the effect of damping overshadows the oscillation. The expression under the radical of Equation (4-20) is positive, and consequently the solution is given directly by Equation (4-21).

2. Forced vibration response

When a system is subjected to an exciting force it is forced to vibrate. The force maybe harmonic (structures subjected to dynamic action of rotating machinery) or of general type ——impulsive, constant, rectangular, triangular, etc. The resulting response of the system to such an excitation is called forced response. The exciting forces and the response of the system are briefly presented below.

(1) Forced harmonic vibration

The system on excitation will vibrate with the same frequency as that of the excitation. For an SDOF system, the exciting force is assumed to be $p_0\sin\omega_e t$ or $p_0\cos\omega_e t$.

For an undamped system, the response is given by

$$x(t) = \frac{x_{st}}{1-r^2}(\sin\omega_e t - r\sin\omega t) \qquad [\text{when } x(0) = 0 \quad \dot{x}(0) = 0] \tag{4-31}$$

For a damped system, the response is given by

$$x(t) = e^{-\xi\omega t}(A\cos\omega_D t + B\sin\omega_D t) + \frac{x_{st}\sin(\omega_e t + \theta)}{\sqrt{(1-r^2)^2 + (2r\xi)^2}} \tag{4-32}$$

Where $x_{st} = \dfrac{p_0}{k}$ is the static deflection of the spring acted upon by the force p_0;

p_0 is the peak amplitude;

ω is the natural frequency of undamped system [Equation (4-13)];

ω_e is the frequency of force or exciting frequency or forcing frequency;

$r = \dfrac{\omega_e}{\omega}$ is the frequency ratio;

ξ is the damping ratio;

$\omega_D = \omega\sqrt{1 - \xi^2}$ is the frequency of the damped system.

The ratio of steady-state amplitude to the static deflection is called dynamic magnification factor D and is given by

$$D = \frac{1}{\sqrt{(1 - r^2) + (2r\xi)^2}} \tag{4-33}$$

(2) Response to general type loading

Often real structures are subjected to forces that are not harmonic. These forces and the corresponding responses are described as follows.

A very large load applied for a very short duration of time that is finite is called an impulsive force. A force $p(t)$ varying arbitrarily with time can be considered as a sequence of infinitesimal short impulses (Figure 4. 9). For a damped system, the response is given by

$$x(t) = \frac{1}{m\omega_D}\int_0^t p(\tau)e^{-\xi\omega(t-\tau)}\sin[\omega_D(t - \tau)]d\tau \tag{4-34}$$

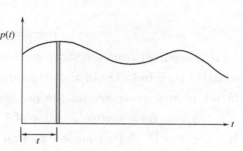

Figure 4. 9   Arbitrary force

In an earthquake, considering $p(t) = -m\ddot{x}_g(t)$, Equation (4-34) can be written as follows:

$$x(t) = -\frac{1}{\omega_D}\int_0^t \ddot{x}_g(\tau)e^{-\xi\omega(t-\tau)}\sin[\omega_D(t - \tau)]d\tau \tag{4-35}$$

This equation is called Duhamel's integral.

During an earthquake, the ground acceleration $\ddot{x}_g$ varies highly irregularly and classical methods of solutions of differential equations of motion are not realistic. Since the response of a structure to the irregular or transient excitation of an earthquake is quite complex, only numerical methods should be used for structural response. It is too difficult for hand calculation in engineering. Then some methods are mentioned for convenience. We introduce one of them called response spectrum analysis.

4. 3   Seismic response calculation of SDOF systems——response spectrum analysis

4. 3. 1   Response spectrum

The structural response to a particular earthquake can be summarized using a response spectrum, which provides valuable information on the potential effects of ground motion on the structure. A response spectrum shows the peak response of an SDOF structure to a particular earthquake, as a function of the natural period and damping ratio of the structure. The main advantage of response spectrum approach is that earthquakes that look quite different when represented in the time domain may actually contain similar frequency contents, and result in broadly similar response spectra. This uniqueness of response spectra makes it useful for a future earthquake.

The concepts of structural dynamics can be used toanalyse the structural response of ground shaking caused by an earthquake. For a linear SDOF system, subjected to ground acceleration, $\ddot{x}_g(t)$, Equation (4-10) can be rewritten as follows:

$$\ddot{x} + \frac{c}{m}\dot{x} + \frac{k}{m}x = -\ddot{x}_g(t) \tag{4-36}$$

or

$$\ddot{x} + 2\xi\omega\dot{x} + \omega^2 x = -\ddot{x}_g(t) \tag{4-37}$$

It is apparent from the above equations that for a given ground motion $\ddot{x}_g(t)$, the deformation response $x(t)$ of the structure depends on the circular frequency ω or period T $(2\pi/\omega)$ of the structure and the damping ratio ξ. Assuming different values of T, say 0.1, 0.2,.... for a particular value of ξ, equation (4-37) is solved for response x (t). Then a curve is plotted among natural period T, deformation x, pseudo-velocity and pseudo-acceleration. The value of deformation when multiplied with ω^2 will give the pseudo-acceleration and when multiplied with ω will give pseudo-velocity. The above process is repeated with different values of ξ, say 0.01, 0.02,... and a set of response curves is thus obtained for the particular selected values of ξ.

For a specific excitation of a simple system having a particular percentage of critical damping, the maximum response is a function of the natural period of vibration of the system. A plot of the maximum response (e.g., relative displacement, absolute displacement, acceleration, or spring force) against the period of vibration, or against the natural frequency of vibration or the circular frequency of vibration, is called a response spectrum.

Seismic design of structures is mostly carried out by equivalent static forces evaluated from the maximum acceleration response of the structure under the expected ground shaking. The seismic effect F can be defined as follows:

$$F = mS_a \tag{4-38}$$

Where　F——the seismic effect;

　　　m——mass;

　　　S_a—— acceleration response spectrum, according to the ground acceleration $\ddot{x}_g$ (t), the circular frequency ω or period T $(2\pi/\omega)$ of the structure and the damping ratio ξ.

Equation (4-38) can be rewritten as follows:

$$F = mS_a = mg \left(\frac{|\ddot{x}_g|_{max}}{g} \right) \left(\frac{S_a}{|\ddot{x}_g|_{max}} \right) = GK\beta = G\alpha \qquad (4-39)$$

In which $K = \dfrac{|\ddot{x}_g|_{max}}{g}$ ——seismic coefficient. It reflects the magnitude of the basic intensity and has nothing to do with the structural dynamic characteristics. The relationship between basic intensity and seismic coefficient is given in Chinese code *The Gode for Seismic Design of Buildings* GB 50011—2010 as follows (Table 4.1).

$\beta = \dfrac{S_a}{|\ddot{x}_g|_{max}}$ ——dynamic coefficient. It is the ratio of the maximum seismic response acceleration of a single point elastic system to the maximum acceleration of ground motion. The maximum dynamic coefficient β_{max} is 2.25 in Chinese code GB 50011—2010.

α ——seismic influence coefficient.

The relationship between basic intensity and seismic coefficient　　　Table 4.1

Basic Intensity	6	7	8	9
Seismic coefficient	0.054	0.107	0.215	0.429

4.3.2   Seismic influence coefficient

The seismic influence coefficient of a building structures shall be determined according to the intensity, site class, design earthquake group, and natural vibration period and damping ratio of structure; the maximum value of its horizontal seismic influence coefficient shall be adopted according to Table 4.2; the characteristic period shall be adopted according to Table 4.3 based on the site class and design earthquake group, and the characteristic period shall be increased by 0.05s to calculate the rare earthquake action.

For the building structures with period larger than 6.0s, the seismic influence coefficient to be adopted shall be studied especially.

Maximum value of horizontal seismic influence coefficient α_{max}　　　Table 4.2

Earthquake effect	Intensity 6	Intensity 7	Intensity 8	Intensity 9
Frequent earthquake	0.04	0.08(0.12)	0.16(0.24)	0.32
Rare earthquake	0.28	0.50(0.72)	0.90(1.20)	1.40

The values in parentheses are used for the zones where the design basic acceleration of ground motion is 0.15g and 0.30g.

Values of characteristic period (s) (T_g)　　　Table 4.3

Design earthquake group	Site class				
	I_0	I_1	II	III	IV
Group 1	0.20	0.25	0.35	0.45	0.65
Group 2	0.25	0.30	0.40	0.55	0.75

Continued

Design earthquake group	Site class				
	I_0	I_1	II	III	IV
Group 3	0. 30	0. 35	0. 45	0. 65	0. 90

The damping adjustment and the form parameters of the seismic influence coefficient curve (Figure 4. 10) of building structure shall meet the following requirements:

(1) Unless otherwise specified, the damping ratio of building structures shall be taken as 0. 05, the damping adjustment coefficient of seismic influence coefficient curve shall be taken as 1. 0 and the form parameter shall meet the following requirements:

1) Linear ascending section, whose period is less than 0. 1s.

2) Horizontal section, whose period is from 0. 1s thought to characteristic period, the maximum value (α_{max}) shall be taken.

3) Curvilinear descending section, whose period is from the characteristic period through to 5 times of the characteristic period, the attenuation index shall be taken as 0. 9.

4) Linear descending section, whose period is from 5 times of the characteristic period through to 6s, the adjustment coefficient for descending slope shall be taken as 0. 02.

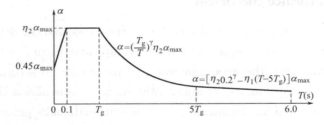

Figure 4. 10　The seismic influence coefficient

In which, α ——the seismic influence coefficient;

α_{max} ——the maximum value of seismic influence coefficient;

η_1 ——the adjustment coefficient for descending slope in the linear decreasing section;

γ ——the attenuation index;

T_g ——the characteristic period;

η_2 ——the damping adjustment coefficient;

T ——the natural vibration period of structure.

(2) If the damping ratio of building structure is not equal to 0. 05 according to the relevant regulations, the damping adjustment coefficient and form parameter of the seismic influence coefficient curve shall meet the following requirements:

1) The attenuation index of curvilinear descending section shall be determined according to thethe following formula:

$$\gamma = 0.9 + \frac{0.05 - \zeta}{0.3 + 6\zeta} \tag{4-40}$$

Where　γ——attenuation index of curvilinear descending section;

　　　　ζ——damping ratio.

　2) The adjustment coefficient for descending slope of attenuation index shall be determined according to the following formula:

$$\eta_1 = 0.02 + \frac{0.05 - \zeta}{4 + 32\zeta} \tag{4-41}$$

Where　η_1——the adjusting factor of slope for the linear decrease section, when it is less than 0, shall equal to 0.

　(3) The damping adjustment coefficient shall be determined according to the following formula:

$$\eta_2 = 1 + \frac{0.05 - \zeta}{0.08 + 1.6\zeta} \tag{4-42}$$

Where　η_2——damping adjustment coefficient, which shall be taken as 0.55 if it is less than 0.55.

In the calculation of earthquake action, the representative value of the gravity load of building shall be taken as the sum of the standard value of deadweight of structure and its components and accessories and the combination value of variable loads. The coefficient for combination value of variable loads shall be adopted according to Table 4.4.

Coefficient for combination values　　　　　Table 4.4

Type of variable load		Coefficient for combination value
Snow load		0.5
Dust load on the roof		0.5
Live load on the roof		Not considered
Live load on floor, calculated according to actual conditions		1.0
Live load on floor, calculated according to equivalent uniform load	Library and archives	0.8
	Other civil buildings	0.5
Gravity of objects suspended by crane	Crane with hard hook	0.3
	Crane with flexible hook	Not considered

The coefficient for combination values of live load on the roof considering the equivalent uniform load of a multi-storey factory shall be taken as 0.8. The crane with hard hook is with high hoisting capacity, and the coefficient for combination value shall be adopted depending on actual conditions.

【Example 4.1】 A one-story reinforced-concrete building is idealized as a massless frame with two columns clamped at the base and a rigid beam supporting a dead load of 1000kN at the beam level (Figure 4.11). The frame is 12m wide and 5m high. The linear stiffness of column $i_c = \dfrac{EI_c}{h} = 2.8 \times 10^4 \text{kN} \cdot \text{m}$. The damping ratio of the building is estimated as 5%. If the building is to be designed for the site Class II in which the seismic precautionary intensity is 8 and design earthquake group is Group 1, determine the lateral seismic effect on the structure in frequent earthquake.

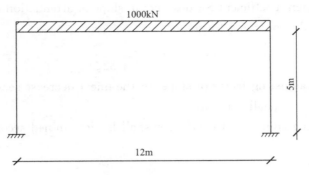

Figure 4. 11   The frame structure

Solution:

The lateral stiffness of the structure $K = 2 \times \left(\dfrac{12i_c}{h^2}\right) = 2 \times \dfrac{12 \times 2.8 \times 10^4}{5^2} = 26880 \text{kN/m}$

The fundamental period $T = 2\pi\sqrt{\dfrac{m}{K}} = 2\pi\sqrt{\dfrac{G}{gK}} = 2 \times 3.1416 \times \sqrt{\dfrac{1000}{9.80 \times 26880}} = 0.387 \text{s}$

Intensity 8, frequent earthquake, $\alpha_{max} = 0.16$ (Table 4.2)

Class II, Group 1, $T_g = 0.35 \text{s}$ (Table 4.3)

Because $T_g < T < 5T_g$

$\alpha = \left(\dfrac{T_g}{T}\right)^{0.9}\alpha_{max} = \left(\dfrac{0.35}{0.387}\right)^{0.9} \times 0.16 = 0.146$

The lateral seismic effect $F = \alpha G = 0.146 \times 1000 = 146 \text{kN}$

4. 4   Seismic response analysis of multiple-degree-of-freedom systems

4. 4. 1   Systems with multiple degrees of freedom

Structures cannot always be modeled as SDOF systems. In fact, structures are continuous systems and possess infinite degrees of freedom. Multi-storey buildings are the most suitable example. A thorough knowledge understanding of concepts discussed for SDOF system are of prime importance. It is because, a structure modeled with MDOF system is transformed to consist of a number of SDOF independent systems and then each one is solved as an SDOF system. The response spectrum of SDOF system is extended to solve MDOF system as well. The MDOF system may be divided into two groups according to their deformation characteristics. In one group, the floor moves only in the horizontal direction and there is no rotation of a horizontal section at the level of floors. Such buildings are referred to as shear buildings. In the other group of structures, the floors move in both rotational and horizontal directions and are referred to as moment-shear buildings.

Considering one of the most instructive and practical type of structure, which involves many degrees of freedom, the multi-storey shear building, the following assumptions are made about the structure:

(1) The total mass of the structure is concentrated at the levels of the floors, although it is distributed throughout the building. This assumption is justified in case of multi-storey buildings where most of the building mass is indeed at the floor levels. This assumption transforms the problem from a structure with infinite degrees of freedom (due to distributed mass), to a structure that has only as many degrees as it has lumped masses at the floor levels. For example, the structure shown in Figure 4.12 has two degrees of freedom.

(2) The girders on the floors are infinitely rigid as compared to the columns and the deformation of the structure is independent of the axial forces present in the columns, this assumption introduces the requirements that the joints between girders and columns are fixed against rotation and the girders remain horizontal during motion.

A building may have any number of bays. It is only as a matter of convenience that we represent the shear buildings solely in terms of a single bay. Further, a shear building can be idealized as a single column (Figure 4.12) having concentrated masses at floor levels, and the columns as massless springs. The stiffness coefficient or spring constant k_j is the force required to produce a unit displacement of the two adjacent floor levels. For a uniform column with the two ends fixed against rotation, the spring constant is $12EI/h^3$, and for a column with one end fixed and the other pinned it is $3EI/h^3$, where E is the modulus of elasticity of the material, I is the moment of inertia, and h is the height of the storey.

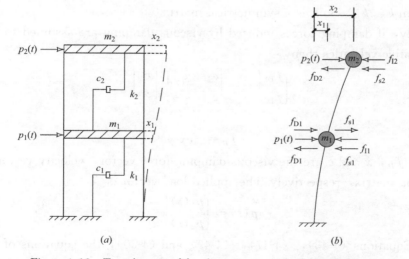

(a)   (b)

Figure 4.12   Two-degrees-of-freedom system under horizontal forces

4.4.2   Dynamic response of multi-storey structure

The equations of motion are developed for a simple MDOF system; a two-storey shear frame is selected to permit easy visualization of elastic, damping, and inertial forces. The following e-quations of motion are obtained for a two-storey shear building [Figure 4. 12 (a)].

$$\begin{bmatrix} f_{I1} \\ f_{I2} \end{bmatrix} + \begin{bmatrix} f_{D1} \\ f_{D2} \end{bmatrix} + \begin{bmatrix} f_{S1} \\ f_{S2} \end{bmatrix} + \begin{bmatrix} p_1(t) \\ p_2(t) \end{bmatrix} = 0 \tag{4-43}$$

The inertial forces in the equations are

$$\begin{bmatrix} f_{I1} \\ f_{I2} \end{bmatrix} = - \begin{bmatrix} m_1 & 0 \\ 0 & m_2 \end{bmatrix} \begin{bmatrix} \ddot{x}_1 \\ \ddot{x}_2 \end{bmatrix} \tag{4-44}$$

or

$$f_I = -m\ddot{x} \tag{4-45}$$

where f_I, $\ddot{x}$ and m are the inertial-force vector, acceleration vector, and mass matrix, respectively.

As shown in Figure 4. 12 (b), the lumped masses are concentrated at floor levels, and the mass matrix is therefore a diagonal matrix. The restoring forces and displacements are related as follows:

$$\begin{bmatrix} f_{S1} \\ f_{S2} \end{bmatrix} = - \begin{bmatrix} k_{11} & k_{12} \\ k_{21} & k_{22} \end{bmatrix} \begin{bmatrix} x_1 \\ x_2 \end{bmatrix} = - \begin{bmatrix} k_1 + k_2 & -k_2 \\ -k_2 & k_2 \end{bmatrix} \begin{bmatrix} x_1 \\ x_2 \end{bmatrix} \tag{4-46}$$

or

$$f_S = -kx \tag{4-47}$$

where f_S, x and k are the elastic-force vector, displacement vector, and stiffness matrix, respectively.

If k_{ij} is the force applied to the ith storey when the jth storey is subjected to a unit displacement, while all other stories remain undisplaced, then by the Maxwell-Betti reciprocal theorem $k_{ij} = k_{ji}$, So k is a symmetrical matrix.

Similarly, if damping forces induced by viscous damping are assumed to be proportional to relative velocities then

$$\begin{bmatrix} f_{D1} \\ f_{D2} \end{bmatrix} = - \begin{bmatrix} c_{11} & c_{12} \\ c_{21} & c_{22} \end{bmatrix} \begin{bmatrix} \dot{x}_1 \\ \dot{x}_2 \end{bmatrix} \tag{4-48}$$

or

$$f_D = -c\dot{x} \tag{4-49}$$

where f_D, $\dot{x}$ and c are the viscous damping-force vector, velocity vector, and viscous damping matrix, respectively. The applied load vector is

$$p(t) = \begin{bmatrix} p_1(t) \\ p_2(t) \end{bmatrix} \tag{4-50}$$

Using Equations (4-45), (4-47), (4-49) and (4-50), the equations of motion for the two-degrees-of-freedom system can be written as

$$m\ddot{x} + c\dot{x} + kx = p(t) \tag{4-51}$$

For an undamped multi-degrees-of-freedom (MDOF) system in free vibration, Equation (4-51) reduces to

$$m\ddot{x} + kx = 0 \tag{4-52}$$

The solution of Equation (4-52) is assumed to be

$$x = X\sin\omega t \tag{4-53}$$

where X represents the vibrational shape (mode shape) of the system. Differentiating Equation (4-53) twice

$$\ddot{x} = -\omega^2 X\sin\omega t \tag{4-54}$$

Substituting Equations (4-53) and (4-54) into Equation (4-52), we can get

$$kX - \omega^2 mX = 0 \tag{4-55}$$

Equation (4-55) is called the frequency equation with respect to the circular frequency ω. For the two-degrees-of-freedom system, we can get

$$\begin{cases} (k_1 + k_2 - \omega^2 m_1)X_1 - k_2 X_2 = 0 \\ -k_2 X_1 + (k_2 - \omega^2 m_2)X_2 = 0 \end{cases} \tag{4-56}$$

For X to have a nontrivial solution, the determinant of Equation (4-56) must be zero.

$$\begin{vmatrix} k_1 + k_2 - \omega^2 m_1 & -k_2 \\ -k_2 & k_2 - \omega^2 m_2 \end{vmatrix} = 0 \tag{4-57}$$

To solve the function, we can get

$$\omega^4 - \left(\frac{k_1 + k_2}{m_1} + \frac{k_2}{m_2}\right)\omega^2 + \frac{k_1 k_2}{m_1 m_2} = 0 \tag{4-58}$$

Four roots can be derived as

$$\omega_{1,2}^2 = \frac{\dfrac{k_1 + k_2}{m_1} + \dfrac{k_2}{m_2} \mp \sqrt{\left(\dfrac{k_1 + k_2}{m_1} + \dfrac{k_2}{m_2}\right)^2 - \dfrac{4k_1 k_2}{m_1 m_2}}}{2} \tag{4-59}$$

The positive roots ω_1 and ω_2 correspond to the first and second natural circular frequencies respectively ($\omega_1 < \omega_2$). By substituting them into Equation (4-56), the ratio of displacements X_2/X_1, is uniquely determined for each ω_1 and ω_2 as shown in Fig4. 13. The modal shapes corresponding to ω_1 and ω_2 are called the first and second mode respectively. As evident from the condition specified by Equation (4-56), only displacement ratio can be obtained. In usual practice, the maximum displacement corresponding to the top or the lowest storey is taken to be unity.

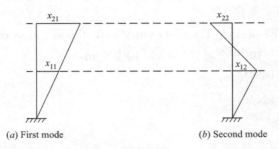

(a) First mode   (b) Second mode

Figure 4. 13   Modal shapes of a two-DOF system

For a system with n degrees of freedom, there will be n natural circular frequencies from Equation (4-55). An n-DOF system can thus vibrate in n different modes, each having a different mode shape and each occurring at a particular natural frequency. The lowest value of ω is called the first natural circular frequency ω_1. The ω are numbered sequentially so that the nth lowest value of ω is the nth natural circular frequency; by substituting it into Equation (4-55), the relative displacements x of the system, which represent the shape of vibration or the modal shape, can be determined.

$$(k - \omega^2 m)X = 0 \tag{4-60}$$

For X to have a nontrivial solution, the determinant of Equation (4-60) must be zero.

$$|k - \omega^2 m| = 0 \tag{4-61}$$

The positive root ω_n corresponds to the nth natural circular frequencies. By substituting them into Equation (4-60), the ratio of displacements X_i/X_1, is uniquely determined for each ω_n and ω_1. The modal shapes corresponding to ω_n is called the nth mode, respectively. In usual practice, the maximum displacement corresponding to the top or the lowest storey is taken to be unity. If a system has n degrees of freedom, then the nth modal shape ϕ_n is written as

$$\phi_n = \begin{bmatrix} \phi_{1n} \\ \phi_{2n} \\ \vdots \\ \phi_{Nn} \end{bmatrix} = \frac{1}{X_{kn}} \begin{bmatrix} X_{1n} \\ X_{1n} \\ \vdots \\ X_{1n} \end{bmatrix} \tag{4-62}$$

Here, X represents the reference component. The square matrix, consisting of n-modal-shape vectors, is called the modal-shape matrix and is expressed as

$$\phi = [\phi_1 \quad \phi_2 \quad \cdots \quad \phi_N] \begin{bmatrix} \phi_{11} & \phi_{12} & \cdots & \phi_{1N} \\ \phi_{21} & \phi_{22} & \cdots & \phi_{2N} \\ \vdots & \vdots & \vdots & \vdots \\ \phi_{N1} & \phi_{N2} & \cdots & \phi_{NN} \end{bmatrix} \tag{4-63}$$

Modal-shape vectors possess an orthogonality relationship for elastic systems.

【Example 4. 2】 Given is a two-storey sheer frame. Whose structural parameters are $m_1 = 60t$, $m_2 = 50t$, $k_1 = 4 \times 10^4 kN/m$, $k_2 = 3 \times 10^4 kN/m$, the frame is 3. 6m high. Determine the natural vibration period of structure and the modal shapes.

Solution:

(1) Calculating the period of natural vibration, the stiffness coefficient is

$k_{11} = k_1 + k_2 = 4 \times 10^4 + 3 \times 10^4 = 7 \times 10^4 kN/m$

$k_{12} = k_{21} = -3 \times 10^4 kN/m$

$k_{22} = k_2 = 3 \times 10^4 kN/m$

$$\begin{vmatrix} k_{11} - m_1\omega^2 & k_{12} \\ k_{21} & k_{22} - m_2\omega^2 \end{vmatrix} = 0$$

$$\begin{vmatrix} 7 \times 10^4 - 60\omega^2 & -3 \times 10^4 \\ -3 \times 10^4 & 3 \times 10^4 - 50\omega^2 \end{vmatrix} = 0$$

$$3\omega^4 - 5300\omega^2 + 12 \times 10^5 = 0$$

$$\omega_1^2 = 266.67 \text{ rad}^2/\text{s}^2, \ \omega_1 = 16.33 \text{rad/s}, \ T_1 = \frac{2\pi}{\omega_1} = \frac{2\pi}{16.33} = 0.385\text{s}$$

$$\omega_2^2 = 1500 \text{ rad}^2/\text{s}^2, \ \omega_2 = 38.73 \text{rad/s}, \ T_2 = \frac{2\pi}{\omega_2} = \frac{2\pi}{38.73} = 0.162\text{s}$$

(2) Calculating the modal shapes

The first modal shape: $\dfrac{x_{12}}{x_{11}} = \dfrac{m_1\omega_1^2 - k_{11}}{k_{12}} = \dfrac{60 \times 267 - 7 \times 10^4}{-3 \times 10^4} = \dfrac{1}{0.556}$

The second modal shape: $\dfrac{x_{22}}{x_{21}} = \dfrac{m_1\omega_2^2 - k_{11}}{k_{12}} = \dfrac{60 \times 1500 - 7 \times 10^4}{-3 \times 10^4} = -\dfrac{1}{1.500}$

$$\sum_{i=1}^{2} m_i x_{1i} x_{2i} = m_1 x_{11} x_{21} + m_2 x_{12} x_{22} = 60 \times 0.556 \times 1.500 + 50 \times 1 \times (-1) = 0$$

The mode shapes are plotted in Figure 4.14.

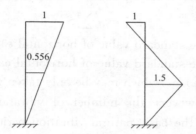

Figure 4.14 Modal shapes of the system

4.5 Seismic response calculation of MDOF systems—— mode-decomposition response spectrum method

If ground acceleration $\ddot{x}_g$, is applied to the structures, then

$$m\ddot{x} + c\dot{x} + kx = -mI\ddot{x}_g \tag{4-64}$$

where m is the lumped mass matrix containing only diagonal terms, k is a banded matrix, and I is a unit vector containing ones corresponding to DOFs in the direction of earthquake load, and zeroes elsewhere.

Similar with the response spectrum method for SDOF system, MDOF system has its response spectrum method for engineering, too. It is called mode-decomposition response spectrum method.

When the mode-decomposition response spectrum method is adopted, the structures not considered with coupling effect shall be calculated for the earthquake action and action effect according to the following requirements:

(1) The standard value of the horizontal earthquake action of a structure at the i th mass point of the j th vibration mode shall be determined according to the following formula:

$$F_{ji} = \alpha_j \gamma_j X_{ji} G_i \quad (i = 1, 2, \cdots, n \quad j = 1, 2, \cdots, m) \tag{4-65}$$

$$\gamma_j = \sum_{i=1}^n X_{ji} G_i / \sum_{i=1}^n X_{ji}^2 G_i \tag{4-66}$$

Where　F_{ji}——standard value of horizontal earthquake action at the ith mass point of the jth vibration mode;

α_j——seismic influence coefficient corresponding to the natural vibration period of the jth vibration mode, which shall be determined according to Article 4.3;

X_{ji}——horizontal relative displacement of the ith mass point of the jth vibration mode;

γ_j——participation coefficient of the jth vibration mode;

G_i——the representative value of the gravity loads concentrated on the ith mass point.

(2) Provided that the period ratio of adjacent vibration modes is less than 0.85, the horizontal earthquake action effects (bending moment, shear force, axial force and deformation) may be determined according to the following formula:

$$S_{Ek} = \sqrt{\sum S_j^2} \tag{4-67}$$

Where　S_{Ek}——effect of the standard value of horizontal earthquake action;

S_j——effect of the standard value of horizontal earthquake action with the jth vibration mode, which may be only taken with the first 2 or 3 vibration modes, however, the number of vibration modes shall be increased properly if the basic natural vibration period is larger than 1.5s or the height-width ratio of the building is larger than 5.

【Example 4.3】 Using mode-decomposition response spectrum method to calculate the standard value of the horizontal earthquake action and shear force of each storey. The building is same as Example 4.2, and to be designed for the site class Ⅱ, in which the seismic precautionary intensity is 8 and design earthquake group is Group 1.

Solution:

(1) Calculating the period of natural vibration

According to Example 4.2,

The first vibration mode: $x_{11} = 0.556$, $x_{12} = 1.000$, $T_1 = 0.385s$

The second vibration mode: $x_{21} = 1.500$, $x_{22} = -1.000$, $T_2 = 0.162s$

(2) Calculating the standard value of horizontal earthquake action

The standard value of horizontal earthquake action of the first vibration mode: $F_{1i} = \alpha_1 \gamma_1 x_{1i} G_i$

$\alpha_{max} = 0.16$(Table 4.2), $T_g = 0.35s$ (Table 4.3)

Because　$T_g < T_1 < 5T_g$,

so　$\alpha_1 = \left(\dfrac{T_g}{T_1}\right)^{0.9} \alpha_{max} = \left(\dfrac{0.35}{0.385}\right)^{0.9} \times 0.16 = 0.147$

$$\gamma_1 = \frac{\sum\limits_{i=1}^n m_i x_{1i}}{\sum\limits_{i=1}^n m_i x_{1i}^2} = \frac{60 \times 0.556 + 50 \times 1}{60 \times 0.556^2 + 50 \times 1^2} = 1.216$$

$F_{11} = 0.147 \times 1.216 \times 0.556 \times 60 \times 9.8 = 58.44 \text{kN/m}$

$F_{12} = 0.147 \times 1.216 \times 1 \times 50 \times 9.8 = 87.59 \text{kN/m}$

The standard value of horizontal earthquake action of the second vibration mode $F_{2i} = \alpha_2 \gamma_2 x_{2i} G_i$

Because $0.1 < T_2 < T_g$, so $\alpha_2 = \alpha_{max} = 0.16$

$$\gamma_2 = \frac{\sum\limits_{i=1}^{n} m_i x_{2i}}{\sum\limits_{i=1}^{n} m_i x_{2i}^2} = \frac{60 \times 1.5 + 50 \times (-1)}{60 \times 1.5^2 + 50 \times (-1)^2} = 0.216$$

$F_{21} = 0.16 \times 0.216 \times 1.5 \times 60 \times 9.8 = 30.48 \text{kN/m}$

$F_{22} = 0.16 \times 0.216 \times (-1) \times 50 \times 9.8 = -16.93 \text{kN/m}$

Calculating the shear force of each floor:

$V_1 = \sqrt{146.03^2 + 13.55^2} = 146.66 \text{kN}$ $\quad$ $V_2 = \sqrt{87.59^2 + 16.93^2} = 89.21 \text{kN}$

4.6 Base shear method

According to the *Code for Seismic Design of Buildings* GB 50011—2010, when some case are satisfied, base shear force method, a approximate method can be adopted.

For structures, which are not higher than 40m, mainly have shear deformation and a rather uniform distribution of mass and rigidity along the height direction, there are some characteristics in vibration:

(1) The displacement response is mainly based on the basic modes.

(2) Basic modes approach straight lines are shown in Figure 4.15.

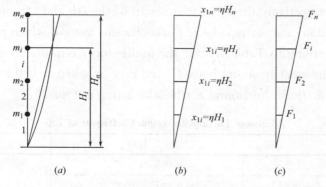

Figure 4.15 Calculation diagram of base shear method

Therefore, the base shear method is suitable for regular buildings with shear deformation less than 40m, such as masonry structures, masonry buildings with internal frame and bottom frame, masonry buildings with seismic walls, single-storey open buildings, single-storey industrial buildings and multi-storey frame structures.

When the base shear force method is adopted, only one degree of freedom may be considered for eachstorey; the standard value of horizontal earthquake action of the structure shall be determined according to the following formula (Figure 4.16).

$$F_{EK} = \alpha_1 G_{eq} \qquad (4-68)$$

$$F_i = \frac{G_i H_i}{\sum_{j=1}^{n} G_j H_j} F_{EK}(1-\delta_n) \quad (i=1,2,\cdots,n) \quad (4-69)$$

$$\Delta F_n = \delta_n F_{EK} \qquad (4-70)$$

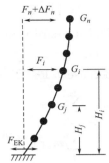

Figure 4.16 Sketch for computation of the horizontal earthquake action

where F_{EK} ——standard value of the total horizontal earthquake action of a structure;

     α_1 ——value of horizontal seismic influence coefficient corresponding to the basic natural vibration period of structure, which shall be determined according to Section 4.3.1; and it should be taken as the maximum value of horizontal seismic influence coefficient for the multi-storey masonry building and the masonry buildings with bottom frame;

     G_{eq} ——total equivalent gravity load of a structure, which shall be taken as the representative value of the total gravity load for single mass point and 85% of the representative value of total gravity load for multi-mass points;

     F_i ——standard value of the horizontal earthquake action at the ith mass point;

     G_i, G_j ——respectively the representative values of the gravity loads concentrated on the ith and jth mass points, which shall be determined according to Table 4.4;

     H_i, H_j ——respectively the calculated height of the ith and jth mass points;

     δ_n ——additional earthquake action coefficient at top, which may be adopted according to Table 4.5 for the multi-storey reinforced concrete houses and the steel structure houses, and may be taken as 0.0 for other buildings;

     ΔF_n ——additional horizontal earthquake action at top.

Additional Earthquake Action Coefficient at Top   **Table 4.5**

T_g (s)	$T_1 > 1.4T_g$	$T_1 \leqslant 1.4T_g$
$T_g \leqslant 0.35$	$0.08T_1 + 0.07$	
$0.35 < T_g \leqslant 0.55$	$0.08T_1 + 0.01$	0.0
$T_g > 0.55$	$0.08T_1 - 0.02$	

Note: T_1 is the basic natural vibration period of the structure.

When the base shear method is used, the earthquake action effects of penthouse, parapet wall and chimney on the roof should be multiplied by an enhancement coefficient of 3; such increase part of effect shall not be transmitted to the lower part of the structure. But the parts connected with this projecting part shall be considered. When modal analysis method is used, the projecting part may be considered as one mass. The enhancement coefficient for the earthquake action effect of projecting skylight frame of a single-storey factory building shall be adopted according to the relevant requirements.

During the seismic check, the horizontal seismic shear force at anystorey of the structure shall comply with the requirements of the following formula:

$$V_{EKi} > \lambda \sum_{j=1}^{n} G_j \tag{4-71}$$

Where V_{EKi} ——storey shear of the ith storey corresponding to the standard value of horizontal earthquake action;

       λ ——coefficient of shear, which shall not be less than the minimum seismic shear force coefficient of storey as specified in Table 4. 6 and also shall be multiplied by an enhancement coefficient of 1. 15 for the weak storey of such structures with vertical irregularity;

       G_j ——representative value of gravity load of the jth storey.

Minimum seismic shear force coefficient value of a storey   Table 4. 6

Type	Intensity 6	Intensity 7	Intensity 8	Intensity 9
Structures with obvious torsion effect or fundamental period less than 3. 5s	0. 008	0. 016(0. 024)	0. 032(0. 048)	0. 064
Structures with fundamental period lager than 5. 0s	0. 006	0. 012(0. 018)	0. 024(0. 036)	0. 048

(1) For the structures with fundamental period between 3. 5s and 5s, the minimum seismic shear force coefficient shall be determined according to interpolation method;

(2) The values in parentheses are used for the zones where the design basic acceleration of ground motion is 0. 15g and 0. 3g respectively.

The horizontal seismic shear force of storeys of a structure shall be distributed according to the following principles:

(1) For the rigid buildings and roof buildings, such as the east-in-situ and monolithic-fabricated concrete floors and roofs, the horizontal seismic shear force should be distributed according to the proportion of equivalent rigidity of lateral-force-resisting components.

(2) For the buildings with flexible floor and roof (such as wood floor and wood roof), the horizontal seismic shear force should be distributed according to the proportion of the representative values of gravity loads in the tributary area of lateral-force-resisting components.

(3) For the buildings with such semi-rigid floor and roof as ordinary prefabricated concrete floor and roof, the horizontal seismic shear force may be taken as the mean value of these above two distribution results.

(4) If the effects of space action, floor deformation, wallelasto-plastic deformation and torsion are considered, the above-mentioned distribution results may be adjusted prop-

erly according to the relevant provisions.

As for the seismic calculation of a structure, the influence of the intersection of base and structure may not be taken into account; as for the reinforced concrete tall buildings that are built in the Class III and IV sites for Intensity 8 and 9 and are adopted with box foundation, raft foundation with good rigidity and the combined pile-box foundation, if the basic natural vibration period of the structure is between 1.2 times and 5 times of the characteristic period and the influence of the intersection of base and structure is considered, then the supposed calculated horizontal seismic shear force of rigid base may be reduced according to the following requirements and the storey drift may be calculated according to the reduced storey shear force.

(1) As for the structures with height-width ratio less than 3, the reduction coefficient for the horizontal seismicshear force of each storey may be calculated according to following formula:

$$\psi = (\frac{T_1}{T_1 + \Delta T})^{0.9} \tag{4-72}$$

Where ψ——reduction coefficient for seismic shear force with consideration of the dynamic interaction of base and structure;

T_1——basic natural vibration period (s) of the structure determined by assumption according to rigid base;

ΔT——additional period with consideration of the dynamic interaction of base and structure, which may be selected according to those specified in Table 4.7.

Additional period (s) Table 4.7

Intensity	Site class	
	III	IV
8	0.08	0.20
9	0.10	0.25

(2) As for the structures with height-width ratio not less than 3, the seismic shear force at bottom shall be reduced according to those specified in Item 1, that at the top of structures shall not be reduced, and that at the middle storeys shall be reduced according to the linear interpolation values.

(3) The horizontal seismicshear force of each storey after reduction shall comply with those specified in Equation (4-71).

【Example 4.4】 Using base shear method to calculate the standard value of the horizontal earthquake action and shear force of each storey. The building is same as Example 4.2 and to be designed for the site class II, in which the seismic precautionary intensity is 8 and design earthquake group is Group 1.

Solution:

According to Example 4.2, $T_1 = 0.385s$, $T_g = 0.35s$(Table 4.3), $\alpha_{max} = 0.16$ (Table

4.2)

$$\alpha_1 = \left(\frac{T_g}{T_1}\right)^{0.9} \alpha_{max} = \left(\frac{0.35}{0.385}\right)^{0.9} \times 0.16 = 0.147$$

$$G_{eq} = \xi \sum_{i=1}^{2} m_i g = 0.85 \times (60 + 50) \times 9.8 = 916 kN$$

$$F_{EK} = \alpha_1 G_{eq} = 0.147 \times 916 = 134.70 kN$$

$$T_1 = 0.385s < 1.4 T_g = 1.4 \times 0.35 = 0.49, \delta_n = 0 \text{ (Table 4.5)}$$

$$\Delta F_2 = \delta_n F_{EK} = 0, F_i = \frac{G_i H_i}{\sum_{i=1}^{n} G_i H_i} F_{EK} (1 - \delta_n)$$

$$\sum_{i=1}^{n} G_i H_i = (60 \times 3.6 + 50 \times 7.2) \times 9.8 = 5644.8 kN \cdot m$$

$$F_1 = \frac{G_1 H_1}{\sum_{i=1}^{n} G_1 H_1} F_{EK}(1 - \delta_n) = \frac{60 \times 9.8 \times 3.6}{5644.8} \times 134.70 \times (1 - 0) = 50.51 kN$$

$$F_2 = \frac{G_2 H_2}{\sum_{i=1}^{n} G_2 H_2} F_{EK}(1 - \delta_n) = \frac{50 \times 9.8 \times 7.2}{5644.8} \times 134.70 \times (1 - 0) = 84.19 kN$$

So $V_1 = F_1 + F_2 + \Delta F_2 = 84.19 + 50.51 + 0 = 134.07 kN$

$V_2 = F_2 + \Delta F_2 = 84.19 + 0 = 84.19 kN$

4.7　Calculation of vertical earthquake action

As for the tall buildings for Intensity 9, the standard value of their vertical earthquake action shall be determined according to the following formula 4-74 (Figure 4.17); the vertical earthquake action effect of storeys may be distributed according to the proportion of the representative values of gravity loads bear by each component and should be multiplied by an enhancement coefficient of 1.5.

$$F_{Evk} = \alpha_{vmax} G_{eq} \tag{4-73}$$

$$F_{vi} = \frac{G_i H_i}{\sum G_j H_j} F_{Evk} \tag{4-74}$$

Where　F_{Evk} ——standard value of the total vertical earthquake action of a structure;

F_{vi} ——standard value of the vertical earthquake action at the ith mass point;

α_{vmax} ——maximum value of vertical seismic influence coefficient, which may be taken as 65% of the maximum value of horizontal seismic influence coefficient;

G_{eq} ——total equivalent gravity load of the structure, which may be taken as 75% of the representative value of its gravity load.

As for such regular flat lattice truss roof with span and length reprehensibly less than those specified in Section 4.8.1 as well as the roof truss, roof transverse beam and bracket

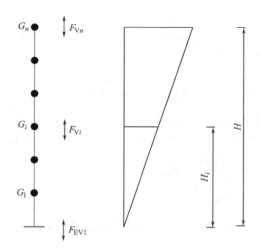

Figure 4. 17 Sketch for computation of the vertical earthquake action

with span larger than 24m, the standard value of their vertical earthquake action should be taken as the product of the representative value of their gravity load and the vertical earthquake action coefficient, in which, the vertical earthquake action coefficient may be adopted according to those specified in Table 4. 8.

Vertical earthquake action coefficient　　　　　　　　**Table 4. 8**

Structure type	Intensity	Site class		
		I	II	III、IV
Flat lattice truss and steel roof truss	8	Not considered (0. 10)	0. 08 (0. 12)	0. 10 (0. 15)
	9	0. 15	0. 15	0. 20
Reinforced concrete roof truss	8	0. 10 (0. 15)	0. 13 (0. 19)	0. 13 (0. 19)
	9	0. 20	0. 25	0. 25

Note: The values in parentheses are used for the zones where the design basic acceleration of ground motion is 0. 30g.

As for the long-cantilever components and the large-span structures not specified in Section 4. 8. 1, the standard value of their vertical earthquake action may be taken 10% and 20% of the representative value of gravity load of this structure or component respectively for Intensity 8 and 9, and also may be taken as 15% of the representative value of gravity load of this structure or component when the design basic acceleration of ground motion is 0. 30g.

The vertical earthquake action of the large-span space structure may also be calculated with the vertical model-decomposition response spectrum method. The vertical seismic influence coefficient hereof may be taken as 65% of the horizontal seismic influence coefficient specified in Section4. 3. 2, however, the characteristic period shall be adopted according to the first design group.

4. 8 General requirements

4. 8. 1 The principle of earthquake action calculation

The earthquake actions of various building structures shall meet the following requirements:

(1) Generally, the horizontal earthquake actions shall be at least considered and checked separately along the two major axial directions of the building structure; and the horizontal earthquake action in each direction shall be bear by the lateral-force-resisting components in this direction.

(2) For the structures having the oblique lateral-force-resisting components, if the intersection angle is greater than 15°, the horizontal earthquake action along the direction of each lateral-force-resisting component shall be calculated respectively.

(3) Structures having obvious asymmetric mass and rigidity distribution, the torsion affects caused by horizontal earthquake actions in two directions shall be considered; for other structures, it is permitted to count in the torsion affects by adjusting the earthquake action effect.

(4) As for the large-span structures and long-cantilevered structures for Intensity 8 or 9 and the tall buildings for Intensity 9, the vertical earthquake action shall be calculated. For building structures adopting seismic isolation for Intensity 8 and 9, the vertical earthquake action shall be calculated according to the relevant regulations.

4. 8. 2 The seismic calculation methods of various building structures

(1) For structures, which are not higher than 40m, mainly have shear deformation and a rather uniform distribution of mass and rigidity along the height direction, or for the structures similar as a single-mass system, the simplified method, such as base shear method, may be adopted.

(2) For building structures other than those as stated in Item 1, the mode-decomposition response spectrum method should be adopted.

(3) The especially irregular buildings, the Category A buildings and the tall buildings belonging to the height range listed in Table 4. 9, shall have the additional calculation under frequent earthquakes by adopting with time-history analysis method.

Height range of buildings adopting time-history analysis method　　　　Table 4. 9

Intensity and site class	Height range of building(m)
Intensity 7, and Class I and II sites for Intensity 8	>100
Class III and IV sites for Intensity 8	>80
Intensity 9	>60

The time-history analysis method is a dynamic analysis method which can directly solve the differential equation of motion of the structure step by step. The time-history a-nalysis method is a relatively time method, which can not only consider the internal tore redistribution after the structure enters the plasticity, but also record the whole process of the structure under the action of a specific seismic wave, which is often not universal.

Exercises

4. 1　What is earthquake action?

4. 2　What are the factors that affect earthquake action?

4. 3　What are the common methods of calculating earthquake action?

4. 4　What is the seismic coefficient? And how to determine it?

4. 5　What is the dynamic coefficient β? And how to determine it?

4. 6　What is the seismic influence factors? And how to determine it?

4. 7　Wnich factors are related to the natural vibration period of the structure?

4. 8　What is time history analysis? What are the advantages of time-history analysis in determining earthquake action?

4. 9　Which structures should consider vertical earthquake action?

4. 10　What is the representative value of gravity load and how to calculate it ?

4. 11　A one-story reinforced-concrete building is idealized as a massless frame with two columns clamped at the base and a rigid beam supporting a dead load of 1000kN at the beam level. The frame is 12m wide and 5m high. The linear stiffness of column $i_c = 2.7 \times 10^4$ kN $\cdot$ m. The damping ratio of the building is estimated as 5%. If the building is to be designed for the site Class II in which the seismic precautionary intensity is 8 and design earthquake group is Group 1, determine the lateral seismic effect on the structure in fre-quent earthquake.

4. 12　Given a two-storey shear frame, simplified as an elastic system with two parti-cles, whose structural parameters are $m_1 = m_2 = 100$t, $k_1 = k_2 = 4 \times 10^4$ kN/m, determine the natural vibration period of structure and the modal shapes.

4. 13　Using mode-decomposition response spectrum method to calculate the standard value of the horizontal earthquake action and shear force in columns of each floor. The building is same as exercise 4. 12, and to be designed for the site class II, in which the seismic precautionary intensity is 8 and design earthquake group is Group 1.

4. 14　Using base shear method to calculate the standard value of the horizontal earth-quake action and shear force of each storey. The building is same as exercise 4. 12, and to be designed for the site class II, in which the seismic precautionary intensity is 8 and de-sign earthquake group is Group 1.

Chapter 5　Seismic Design of Reinforced Concrete Structures

5.1　Introduction

The main load-carrying members of reinforced concrete structure are constructed by reinforced concrete that refers to a kind of composite material which can improve the mechanical properties of concrete by adding steel mesh, steel plate or fiber into concrete. It has the advantages of ruggedness, durability, good fire resistance, saving steel, local materials and low cost compared with steel structure, and the disadvantages of high density and poor crack resistance.

Reinforced concrete structure types generally include frame structure suitable for office buildings, teaching buildings and so on, seismic wall structure for residential buildings, frame-seismic wall structure also for residential buildings, partial frame-support seismic wall structure for underground garage and tube structure for super high-rise buildings above 30 stories.

Reinforced concrete frame structure is a bearing system mainly consisting of beams and columns that resists horizontal loads and vertical loads during its serviceability. It has the advantage of flexible arrangement of space, the filler walls are mainly used to separate the space, so it can be often used in shopping malls, industrial workshops, auditoriums, residential buildings, office buildings, hospitals buildings and school buildings.

Under the horizontal seismic action or wind load, the internal forces of frame structure members near the bottom floor and the lateral displacement of the building increase sharply with the increase of the building height. Therefore, when the building exceeds a certain height, its lateral stiffness will decreases significantly, if the frame structure is still adopted, the cross sections of the frame beam and column will be very large. Thus, the cost of the building will not only increase, but also the building area will decrease. In this case, the seismic wall structure or frame-seismic wall structure is usually adopted.

The seismic wall structure refers to the structure, in which the main load-bearing members are all the reinforced concrete walls, and its lateral stiffness is very large in the plane of the walls, and the out-of-plane stiffness of the walls is quite small and often can be neglected in the calculation analysis of structure. It has the advantage of the larger lateral stiffness than frame, so its height limit is greater than the one of frame, and it has the disadvantage of inflexible arrangement of space and larger seismic action. This structure type is often used in the high-rise apartment buildings.

The frame-seismic wall structure is that a certain number of seismic walls are arranged in a frame structure to make it have the considerable lateral stiffness to resist the deformation under wind load and seismic action, and because the seismic walls have the larger lateral stiffness than the columns in frame, most of the seismic shear force is bear by the seismic walls and the rest by the columns in frame, and the frame mainly bears the gravity load. The frame-seismic wall structure combines the frame's advantage of flexible arrangement of space and the seismic-wall's advantage of the larger lateral stiffness, so this structure type is widely used in the high-rise buildings.

Compared with the masonry buildings, the frame buildings, the seismic wall buildings and the frame-seismic wall buildings have the higher bearing capacity, ductility and integrity, and they are widely used in the high seismic intensity areas due to the better seismic performance.

5.2　Seismic damage and analysis

It is known that knowledge comes from practice and guides practice, and only a correct understanding can play a positive role in guiding practice, so earthquake disaster and engineering experience are the main sources of seismic conceptual design of building. Seismic damage always makes structure engineers identify the weak location of structure and causes the revision of seismic design codes.

5.2.1　Damage of frame components

The seismic damage of frame beams and columns has mainly found in the joints of beams and columns, the seismic damage of columns is more severe than that of beams; the seismic damage of column top is more severe than that of column bottom; the seismic damage of corner columns is more severe than that of inner columns; and the seismic damage of short columns is more severe than that of ordinary columns.

The typical damage characteristics of reinforced concrete frame members are shown in Table 5.1.

Seismic damages of frame members　　　　　　　　　　Table 5.1

Member	Location	Characteristics	Reasons
Column	Top	Horizontal or diagonal cracks	Greater internal forces at the top; insufficient stirrups or poor anchorage
		Concrete crushing and collapse	
		Longitudinal steel bars bent like a lantern	
		Broken stirrups	
	Bottom	Horizontal cracks	Greater internal forces at the bottom
		Concrete spalling	
		Reinforcement yielding	
	Other location	Diagonal cracks	Lack of sufficient shear capacity

Member	Location	Characteristics	Reasons
Beam	Two ends	Vertical cracks	Repeated cycle seismic action
		Diagonal cracks	Lack of stirrups
Joint	—	Diagonal cracks	Lack of stirrups; large and complex stresses; loose concrete due to limited space
		Concrete crushing	
		Longitudinal steel bars bent like a lantern	

5.2.2   Damage of seismic walls

The seismic walls can bear the internal forces caused by various loads, and can effectively control the horizontal force of the structure. The shear damage of seismic walls is primarily located in the coupling beam with smaller span-depth ratio, in which X-shaped cracks usually occur. Horizontal cracks propagate at the top or bottom of seismic walls, especially when the walls are thin or the horizontal steel bars in them are insufficient.

5.2.3   Damage of infilled walls

The brick-filled walls of frame structure are seriously damaged, and the X-shaped cracks or even collapse often appear in these walls when the seismic intensity reaches 7 degrees, the reason is that the inter-story displacements of the frame are large under seismic action, the infilled walls try to prevent the lateral displacement of frame, and the ultimate deformation of brick masonry is small.

The deformation of frame structure is the shear type, and the displacements at the lower stories are larger due to the larger story seismic shear forces, so the seismic damage of infilled walls is severe in the middle and lower stories of building. The deformation of frame seismic wall structure is close to the bending type, and the displacements at the upper stories are large, so the seismic damage of infilled walls in the upper stories of buildings is more severe.

5.2.4   Other damage

The high-rise buildings located at soft soil are prone to severe damage due to the resonance that the natural period of structure is close to that of soil. For example, in the 1976 Tangshan M7.8 earthquake, the collapse occurred above seven stories of a 13-story frame structure at the Tanggu district of Tianjin which is 70 kilometers from the epicenter.

When the width of seismic joint is too small, the structures collide with each other easily. For example, there is a 150 mm seismic joint between the east and west sections of Tianjin Friendship Hotel, many bricks fall from the seismic joint due to collision during the Tangshan earthquake.

The structures on soft soil or a liquefiable soil layer are prone to the whole inclination or even collapse in earthquakes because of the uneven settlement of base.

5.3 Structure system and seismic grade

5.3.1 Selection of structure system

According to the investigation of earthquake damage and engineering experience at home and abroad, the maximum heights of cast-in-situ reinforced concrete buildings are related to the structure type, fortification intensity and site category in order to meet the requirements of safety and economy, and are given in Table 5.2 according to the Chinese code, in which the height of building refers to the height between the outdoor ground and the top of the main roof slab and excludes the partially protruding roof. As for structures that are irregular horizontally and vertically, the applicable maximum height should be reduced properly.

Applicable maximum height of cast-in-situ reinforced concrete buildings (m)   Table 5.2

Structure type	Seismic fortification intensity				
	6	7	8(0.2g)	8(0.3g)	9
Frame structure	60	50	40	35	24
Frame-seismic wall	130	120	100	80	50
Seismic wall	140	120	100	80	60
Partial frame-support seismic wall	120	100	80	50	Inapplicable
Frame-core-tube	150	130	100	90	70
Tube-in-tube	180	150	120	100	80
Slab-column-seismic wall	80	70	55	40	Inapplicable

5.3.2 Seismic grade of reinforced concrete structures

The reinforced concrete buildings shall be designed with different seismic grades according to the fortification category, fortification intensity, structure type and building height, and also shall meet the requirements of the corresponding calculation and structural measures. The seismic grade of Category C buildings shall be adopted according to the Table 5.3 which only shows the seismic grades of the frame structures, frame-seismic wall structures and seismic wall structures, and that of other structure types can be looked up in the *Code for Seismic Design of Buildings* GB 50011—2010. The seismic grades of the Category A, B, and D buildings should be specified according to the seismic fortification intensity which the seismic measures in the seismic fortification standards require.

The large-span frame in Table 5.3 refers to a frame whose span is over than 18m. The determination of the seismic grade of reinforced concrete building still should meet some other requirements which can be consulted from the Chinese code GB 50011—2010 for seismic design of buildings.

Seismic grades of cast-in-situ reinforced concrete buildings (m)　　　Table 5.3

Structure type		Fortification intensity								
		6		7		8		9		
Frame	Height(m)	≤24	>24	≤24	>24	≤24	>24	≤24		
	Frame	Ⅳ	Ⅲ	Ⅲ	Ⅱ	Ⅱ	Ⅰ	Ⅰ		
	Large-span frame	Ⅲ		Ⅱ		Ⅰ		Ⅰ		
Frame-seismic wall	Height(m)	≤60	>60	≤24	25~60	>60	≤24	25~60	>60	≤24	25~50
	Frame	Ⅳ	Ⅲ	Ⅳ	Ⅲ	Ⅱ	Ⅲ	Ⅱ	Ⅰ	Ⅱ	Ⅰ
	Seismic wall	Ⅲ		Ⅲ		Ⅱ		Ⅱ	Ⅰ	Ⅰ	
Seismic wall	Height(m)	≤80	>80	≤24	25~80	>80	≤24	25~80	>80	≤24	25~60
	Seismic wall	Ⅳ	Ⅲ	Ⅳ	Ⅲ	Ⅱ	Ⅲ	Ⅱ	Ⅰ	Ⅱ	Ⅰ

5.3.3　Seismic joint

The reinforced concrete buildings, requiring the arrangement of seismic joints, shall meet the following requirements.

For the buildings with frame structures, including frame structure with a few seismic walls, the width of seismic joint shall not be less than 100mm when the height of building does not exceed 15m; if not, the width of seismic joint should be increased by 20mm respectively when the building height is increased by every 5m, 4m, 3m and 2m for the seismic fortification Intensity 6, 7, 8 and 9 respectively.

For the buildings with frame-seismic wall structures and the buildings with seismic wall structures, the width of seismic joints shall not be less than 0.7 times and 0.5 times the width of the seismic joints of buildings with frame structures respectively, and the width for both two kinds of buildings should not be less than 100mm.

The width of seismic joint shall be specified according to the structure type requiring wider seismic joint and the smaller building height when the structures on the two sides of seismic joint are of different types.

For the buildings with frame structures for the seismic fortification Intensity 8 and 9, the stirrups of the frame columns on both sides of the seismic joint shall be densified along the total height of the building. The internal forces of frame members shall be calculated and adopted according to the unfavorable situations of two kinds of calculation models containing the anti-collision walls or not.

5.4　Seismic design of reinforced concrete frame

5.4.1　Calculation of lateral seismic action

Generally, the lateral seismic actions should be at least considered and checked separately along the two major axial directions of the building structure, and the lateral seismic action along each direction should be bear by the lateral-force-resisting members in this direction.

The base shear method may be adopted for the building structures with the height less than 40m that mainly has the shear deformation and a rather uniform distribution of both mass and rigidity vertically, or for the building structures simplified as a single degree of freedom system, otherwise, the response spectrum method or the time history method should be used.

5. 4. 2　Approximate calculation of basic natural vibration period

The basic natural vibration period of the structure must be known before calculating the lateral seismic action by the base shear method. Generally, we cannot always obtain the exact periods because of the inevitable difference between the computation model and the actual structure; however, for structure engineers, the approximate value of the basic natural vibration period can already satisfy the structure design. There are several approaches for approximately estimating the basic natural vibration period such as the Rayleigh method, the top-drift method, the equivalent mass method and numerical dynamic analysis.

For the multi-story reinforced concrete frame, the lateral story displacements are calculated easily, so the Rayleigh method is often used to calculate the basic natural vibration period, and the top-drift method can be adopted usually for the high-rise reinforced concrete frame and frame-seismic wall structure.

1. The Rayleigh method

Based on the energy conservation principle, the Rayleigh method is used for the approximate calculation of the basic vibration frequency of structure. When an elastic system without damping vibrates freely, the total energy of system will be not changeable at any time, and the total energy is always equal to the sum of strain energy and kinetic energy.

Suppose that the length and distributed mass of a simply-supported beam could be denoted as l and $\overline{m}$ respectively, and the displacement of simply supported beam with free vibration is as follows:

$$y(x,t) = Y(x)\sin(\omega t + \theta) \tag{5-1}$$

Where $Y(x)$ is the displacement amplitude at the location x, ω is the circular frequency of free vibration, t is time, θ is phase angle.

It's know that the first-order derivative of displacement to time is velocity, that is

$$\dot{y}(x,t) = \omega Y(x)\cos(\omega t + \theta) \tag{5-2}$$

Where one dot above y denotes the first-order derivative of displacement to time.

The bending strain energy is as follows:

$$U = \frac{1}{2}\int_0^l EI\left(\frac{\partial^2 y}{\partial x^2}\right)^2 dx = \frac{1}{2}\sin^2(\omega t + \theta)\int_0^l EI[Y''(x)]^2 dx \tag{5-3}$$

Where the two apostrophes in the upper right of Y denote the second-order derivative of displacement amplitude to the location.

The maximum of U is as follows:

$$U_{max} = \frac{1}{2} \int_0^l EI [Y''(x)]^2 dx \qquad (5-4)$$

The kinetic energy is as follows:

$$T = \frac{1}{2} \int_0^l \overline{m} (\frac{\partial y}{\partial t})^2 dx = \frac{1}{2} \omega^2 \cos^2(\omega t + \theta) \int_0^l \overline{m} [Y(x)]^2 dx \qquad (5-5)$$

The maximum of T is as follows

$$T_{max} = \frac{1}{2} \omega^2 \int_0^l \overline{m} [Y(x)]^2 dx \qquad (5-6)$$

If $\sin(\omega t + \theta)$ is zero, the displacement and the strain energy are zero, the velocity and the kinetic energy are maximum, and the total energy of system is equal to T_{max}.

If $\cos(\omega t + \theta)$ is zero, the velocity and kinetic energy are zero, the displacement and the strain energy are maximum, and the total energy of system is equal to U_{max}.

According to the energy conservation principle, the following expression could be derived easily

$$T_{max} = U_{max} \qquad (5-7)$$

The circular frequency of free vibration can be obtained further as follows:

$$\omega^2 = \frac{\int_0^l EI [Y''(x)]^2 dx}{\int_0^l \overline{m} [Y(x)]^2 dx} \qquad (5-8)$$

If the lumped masses $m_i (i = 1, 2, \cdots, n)$ are located on the beam, the frequency formula can be modified as follows:

$$\omega^2 = \frac{\int_0^l EI [Y''(x)]^2 dx}{\int_0^l \overline{m} [Y(x)]^2 dx + \sum_{i=1}^{n} m_i Y_i^2} \qquad (5-9)$$

Where Y_i is the displacement amplitude of lumped mass m_i.

The vibration modes are generally unknowable, so $Y(x)$ need to be supposed. If $Y(x)$ is identical with a certain vibration mode, the accurate frequency can be obtained.

The deformation curve of simply-supported beam under the distribution load $q(x)$ (structure self-weight, etc.) could be considered approximately as the basic vibration mode, the strain energy is equal to the work done by the distribution load $q(x)$, the frequency formula can be obtained further as follows:

$$\omega^2 = \frac{\int_0^l q(x) Y(x) dx}{\int_0^l \overline{m} [Y(x)]^2 dx + \sum_{i=1}^{n} m_i Y_i^2} \qquad (5-10)$$

If the deformation curve of simply-supported beam under the self-weight load could be considered as the approximate value of $Y(x)$, the basic frequency formulas could be modified further as follows:

$$\omega^2 = \frac{\int_0^l \overline{m}gY(x)\,\mathrm{d}x + \sum_{i=1}^{n} m_i gY_i}{\int_0^l \overline{m}[Y(x)]^2\,\mathrm{d}x + \sum_{i=1}^{n} m_i Y_i^2} \tag{5-11}$$

Where g is the gravity acceleration.

According to the relationship between circular frequency and period, the basic vibration period could be given as follows:

$$T_1 = \frac{2\pi}{\omega} = 2\pi \sqrt{\frac{\int_0^l \overline{m}[Y(x)]^2\,\mathrm{d}x + \sum_{i=1}^{n} m_i Y_i^2}{\int_0^l \overline{m}gY(x)\,\mathrm{d}x + \sum_{i=1}^{n} m_i gY_i}} \tag{5-12}$$

It needs to be mentioned that if the horizontal vibration of structure is considered, the action direction of gravity should be in accordance with the horizontal direction, that is to say, the deformation curve of structure under the assumed load, whose direction is horizontal and value is equal to gravity, could be approximately considered as the basic vibration mode along the horizontal direction. For example, when a four-storey frame vibrates laterally, the structure can be simplified as the 4-degree of freedom system with lumped mass G_i, $i=1$, 2, 3, 4, and the lateral calculation diagram is shown in Figure 5.1.

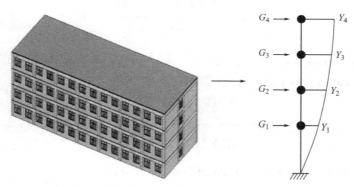

Figure 5.1   The multi-degree of freedom
system with lumped masses

The Rayleigh method is used to calculate the basic vibration period of frame structure. Because the multi-degree of freedom system only has the lumped masses, the formula (5-12) could be simplified further as follows:

$$T_1 = 2\pi \sqrt{\frac{\sum_{i=1}^{n} m_i Y_i^2}{\sum_{i=1}^{n} m_i gY_i}} \tag{5-13}$$

The reduction factor φ_T should be multiplied in the formula (5-13) to consider the influence of non-bear wall stiffness on the structure period, and substituting $\pi^2/g \approx 1$ and

$G_i = m_i g$ into the formula (5-13)

$$T_1 = 2\varphi_T \sqrt{\frac{\sum\limits_{i=1}^{n} G_i Y_i^2}{\sum\limits_{i=1}^{n} G_i Y_i}} \qquad (5\text{-}14)$$

It is important to note that the unit of Y_i must be meter. The reduction factor φ_T is between 0.6 and 0.7 for the civil frame structure.

The calculation procedure of basic vibration period of frame structure is as follows:

(1) The ith story shear forces under assumed horizontal gravity load:

$$V_i = \sum_{j=i}^{n} G_j \qquad (5\text{-}15)$$

(2) The ith story shifts:

$$\delta_i = \frac{V_i}{K_i} \qquad (5\text{-}16)$$

Where K_i is the ith story lateral stiffness.

(3) The shifts of ith lumped masses relative to ground:

$$Y_i = \sum_{j=1}^{i} \delta_j \qquad (5\text{-}17)$$

(4) Substituting G_i and Y_i into the formula (5-14), the basic vibration period will be obtained.

【Example 5.1】 A two-story frame structure has three spans along the lateral direction and seven spans along the longitudinal direction. The representative values of gravity load are $G_1 = 9420\text{kN}$ on the first floor and $G_2 = 7750\text{kN}$ on the second floor, respectively. The lateral stiffness of each column on the first floor and on the second floor is $k_1 = 8500\text{kN/m}$ and $k_2 = 13500\text{kN/m}$, respectively. What is the lateral basic vibration period of structure? ($\varphi_T = 0.7$)

Solution:

(1) The story shear forces under assumed horizontal gravity load:

$V_2 = G_2 = 7750\text{kN}$

$V_1 = G_1 + G_2 = 17170\text{kN}$

(2) The story shifts:

$$\delta_2 = \frac{V_2}{K_2} = \frac{V_2}{(3+1)(7+1)k_2} = 0.018\text{m}$$

$$\delta_1 = \frac{V_1}{K_1} = \frac{V_1}{(3+1)(7+1)k_1} = 0.063\text{m}$$

(3) The shifts of lumped masses relative to ground:

$Y_2 = \delta_1 + \delta_2 = 0.081\text{m}$

$Y_1 = \delta_1 = 0.063\text{m}$

(4) Using the formula (5-14), $T_1 = 0.37\text{s}$.

2. The top-drift method

As its name implies, the basic natural vibration period of structure is calculated directly by the top drift of structure subject to the gravity load acting horizontally (Figure 5.2).

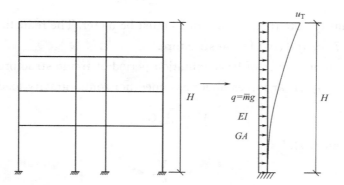

Figure 5.2   The top-drift method

If the structure could be regarded as the shear type rod, the equation of lateral free motion without damping for the rod is as follows:

$$\overline{m}\frac{\partial^2 w(z,t)}{\partial t^2} - \frac{GA}{\mu}\frac{\partial^2 w(z,t)}{\partial z^2} = 0 \tag{5-18}$$

Where GA is the shear stiffness, μ is the numerical factor depending on the shape of the bar cross section, $\overline{m}$ is the uniform distribution mass. By the variable separation approach, the basic natural vibration period could be obtained as follows:

$$T_1 = 4H\sqrt{\frac{\mu\overline{m}}{GA}} \tag{5-19}$$

Where H is the height of structure.

From structure mechanics, the top drift of structure under the uniform distribution load $q = \overline{m}g$ with lateral direction is as follows:

$$u_s = \frac{\mu q H^2}{2GA} \tag{5-20}$$

Substituting formula (5-20) into formula (5-19), the basic natural vibration period is as follows:

$$T_1 = 1.8\sqrt{u_s} \tag{5-21}$$

If the structure could be regarded as the bending type rod, the equation of lateral free motion without damping for the rod is as follows:

$$\overline{m}\frac{\partial^2 w(z,t)}{\partial t^2} + EI\frac{\partial^4 w(z,t)}{\partial z^4} = 0 \tag{5-22}$$

Where EI is the bending stiffness, $\overline{m}$ is the uniform distribution mass. By the variable separation approach, the basic natural vibration period could be obtained as follows:

$$T_1 = 1.787H^2\sqrt{\frac{\overline{m}}{EI}} \tag{5-23}$$

From structure mechanics, the top drift of structure under the uniform distribution load $q = \overline{m}g$ with lateral direction is as follows:

$$u_b = \frac{qH^4}{8EI} \tag{5-24}$$

Substituting formula (5-24) into formula (5-23), the basic natural vibration period is as follows:

$$T_1 = 1.6\sqrt{u_b} \tag{5-25}$$

Generally, the practical structure such as frame structure, seismic wall and frame-seismic wall should be regarded as both the bending-shear type rod, and the influence of non-bear wall stiffness on the structure period should be considered, the basic natural vibration period can be calculated by the following formula:

$$T_1 = 1.7\varphi_T\sqrt{u_T} \tag{5-26}$$

Where u_T is the top drift of structure under the representative value of gravity loads acting horizontally, its unit must be meter, and the reduction factor φ_T is $0.6 \sim 0.7$ for frame structure, $0.7 \sim 0.8$ for frame-seismic wall and $0.8 \sim 1.0$ for seismic wall.

In the Example 5.1, substituting $u_T = 0.081\text{m}$ and $\varphi_T = 0.7$ into (5-26), $T_1 = 0.34\text{s}$.

5.4.3   Internal force and drift of frame structure

The multi-story reinforced concrete frame structure is statically indeterminate structures of high order, so the approximate methods are used usually for the calculation of practical engineering structure.

1. D-value method for horizontal load

The slope-deflection equation of a straight rod with equal cross-section is as follows:

$$\begin{Bmatrix} M_A \\ M_B \\ V_{AB} \end{Bmatrix} = \begin{bmatrix} 4i & 2i & -\dfrac{6i}{l} \\ 2i & 4i & -\dfrac{6i}{l} \\ -\dfrac{6i}{l} & -\dfrac{6i}{l} & -\dfrac{12i}{l^2} \end{bmatrix} \begin{Bmatrix} \theta_A \\ \theta_B \\ \Delta_{AB} \end{Bmatrix} \tag{5-27}$$

Where M_A and M_B are the bending moments of rod ends, θ_A and θ_B are the rotation angles of bar ends, V_{AB} is the bar shear force, Δ_{AB} is the relative displacement of rod ends along the direction perpendicular to the rod axis, $i = EI/l$ and l are the linear stiffness and length of rod, respectively. E is elastic modulus, I is the inertia moment of cross section.

(1) Lateral stiffness of non-bottom column

The assumptions are made as follows (Figure 5.3):

1) The end rotation angles of both column AB and the rods adjacent to column AB are θ.

2) The linear stiffness of both column AB and its adjacent upper and lower columns are i_c, and their chord rotation angles are ϕ.

According to the bending moment balance of joint A and joint B, respectively, the relationship between θ and ϕ could be obtained as follows:

$$\theta = \frac{2}{2+\overline{K}}\phi \qquad (5\text{-}28)$$

Where $\overline{K} = (i_{b1} + i_{b2} + i_{b3} + i_{b4})/(2i_c)$ is called the stiffness ratio of beam to column, i_{b1}, i_{b2}, i_{b3} and i_{b4} are the linear stiffness of beams, i_c is the linear stiffness of columns.

According to the formula (5-27), the shear force of column AB is as follows:

$$V_{AB} = \frac{12i_c}{h_{AB}}(\phi - \theta) \qquad (5\text{-}29)$$

Substituting formula (5-28) into formula (5-29), the lateral stiffness of column AB could be obtained as follows:

$$D_{AB} = \alpha \frac{12i_c}{h_{AB}^2} \qquad (5\text{-}30)$$

Where $\alpha = \overline{K}/(2+\overline{K})$, is called the modified factor of stiffness considering the rotation of frame joints.

(2) Lateral stiffness of bottom column

According to the bending moment balance of joint J in Figure 5.3, the lateral stiffness of column JK could be obtained as follows:

$$D_{JK} = \alpha \frac{12i_c}{h_{JK}^2} \qquad (5\text{-}31)$$

Where $\alpha = (2+3\overline{K})/(5+3\overline{K})$, $\overline{K} = (i_{b5} + i_{b6})/i_c$ is called the stiffness ratio of beam to column, i_{b5} and i_{b6} are the linear stiffness of beams, i_c is the linear stiffness of columns. In the engineering calculation, the modified factor α could be approximately $(0.5+\overline{K})/(2+\overline{K})$.

It must be mentioned that the effect of slabs on beam stiffness should be considered, that is, the slabs as the beam flanges work with beams. In the engineering application, for simplicity, the inertia moment of beam with rectangle cross section should be calculated first, and then are multiplied by the amplification factors in Table 5.4, in which the interior frame is the frame whose beams have slabs at both sides, however, the beams of exterior frame have slab only at one side.

Amplification factors of the inertia moment of frame beam   Table 5.4

Construction type	Interior frame	Exterior frame
Cast-in-situ beam slab	2.0	1.5
Precast composite beam slab	1.5	1.2
Precast beam slab	1.0	1.0

(3) Column contra-flexure point

The height of column contra-flexure point can be calculated by the following formula:

$$h' = (y_0 + y_1 + y_2 + y_3)h \qquad (5\text{-}32)$$

Where y_0 is called the height ratio of standard contra-flexure point, and could be seen in Table 5.5 by the story number m of frame, the number n of storey on which the column is located and the stiffness ratio $\overline{K}$ of beam to column. The adjustment coefficients y_1 by the stiffness ratio α_1 of beams at column ends and y_2 (y_3) by the height ratio α_2 (α_3) of adjacent columns could be seen in Table 5.6 and Table 5.7, respectively.

The height ratio y_0 of standard contra-flexure point for inverted triangular loads　　Table 5.5

m	$\overline{K}$ / n	0.1	0.2	0.3	0.4	0.5	0.6	0.7	0.8	0.9	1.0	2.0	3.0	4.0	5.0
1	1	0.80	0.75	0.70	0.65	0.65	0.60	0.60	0.60	0.60	0.55	0.55	0.55	0.55	0.55
2	2	0.50	0.45	0.40	0.40	0.40	0.40	0.40	0.40	0.40	0.45	0.45	0.45	0.45	0.50
	1	1.00	0.85	0.75	0.70	0.65	0.65	0.65	0.65	0.60	0.60	0.55	0.55	0.55	0.55
3	3	0.25	0.25	0.25	0.30	0.30	0.35	0.35	0.35	0.40	0.40	0.45	0.45	0.45	0.50
	2	0.60	0.50	0.50	0.50	0.50	0.45	0.45	0.45	0.45	0.45	0.50	0.50	0.50	0.50
	1	1.15	0.90	0.80	0.75	0.75	0.70	0.70	0.65	0.65	0.65	0.55	0.55	0.55	0.55
4	4	0.10	0.15	0.20	0.25	0.30	0.35	0.35	0.35	0.35	0.40	0.45	0.45	0.45	0.45
	3	0.35	0.35	0.35	0.40	0.40	0.40	0.40	0.45	0.40	0.45	0.45	0.50	0.50	0.50
	2	0.70	0.60	0.55	0.50	0.50	0.50	0.50	0.50	0.50	0.50	0.50	0.50	0.50	0.50
	1	1.20	0.95	0.85	0.80	0.75	0.70	0.70	0.65	0.65	0.65	0.55	0.55	0.55	0.55
5	5	−0.05	0.10	0.20	0.25	0.30	0.30	0.35	0.35	0.35	0.35	0.40	0.45	0.45	0.45
	4	0.20	0.25	0.35	0.35	0.40	0.40	0.40	0.40	0.45	0.45	0.45	0.50	0.50	0.50
	3	0.45	0.40	0.45	0.45	0.45	0.45	0.45	0.45	0.45	0.50	0.50	0.50	0.50	0.50
	2	0.75	0.60	0.55	0.55	0.55	0.50	0.50	0.50	0.50	0.50	0.50	0.50	0.50	0.50
	1	1.30	1.00	0.85	0.80	0.75	0.70	0.70	0.65	0.65	0.65	0.60	0.55	0.55	0.55
6	6	−0.15	0.05	0.15	0.20	0.25	0.30	0.30	0.35	0.35	0.35	0.40	0.45	0.45	0.45
	5	0.10	0.25	0.30	0.35	0.35	0.40	0.40	0.40	0.35	0.45	0.45	0.50	0.50	0.50
	4	0.30	0.35	0.40	0.40	0.45	0.45	0.45	0.45	0.45	0.45	0.50	0.50	0.50	0.50
	3	0.50	0.45	0.45	0.45	0.45	0.45	0.45	0.45	0.45	0.50	0.50	0.50	0.50	0.50
	2	0.80	0.65	0.55	0.55	0.55	0.55	0.50	0.50	0.50	0.50	0.50	0.50	0.50	0.50
	1	1.30	1.00	0.85	0.80	0.75	0.70	0.70	0.65	0.65	0.65	0.60	0.55	0.55	0.55

y_1 by the stiffness ratio α_1 of beams at column ends　　Table 5.6

α_1 / $\overline{K}$	0.1	0.2	0.3	0.4	0.5	0.6	0.7	0.8	0.9	1.0	2.0	3.0	4.0	5.0
0.4	0.55	0.40	0.30	0.25	0.20	0.20	0.20	0.15	0.15	0.15	0.05	0.05	0.05	0.05
0.5	0.45	0.30	0.20	0.20	0.15	0.15	0.15	0.10	0.10	0.10	0.05	0.05	0.05	0.05
0.6	0.30	0.20	0.15	0.15	0.10	0.10	0.10	0.10	0.05	0.05	0.05	0.05	0	0
0.7	0.20	0.15	0.10	0.10	0.10	0.10	0.05	0.05	0.05	0.05	0.05	0	0	0
0.8	0.15	0.10	0.05	0.05	0.05	0.05	0.05	0.05	0.05	0	0	0	0	0
0.9	0.05	0.05	0.05	0.05	0	0	0	0	0	0	0	0	0	0

Note: if $i_{b1} + i_{b2} > i_{b3} + i_{b4}$, $\alpha_1 = \dfrac{i_{b3} + i_{b4}}{i_{b1} + i_{b2}}$ (Figure 5.4), y_1 is negative.

y_2 and y_3 by the height ratio α_2 (α_3) of adjacent columns   Table 5.7

α_2 \ $\overline{K}$ \ α_3		0.1	0.2	0.3	0.4	0.5	0.6	0.7	0.8	0.9	1.0	2.0	3.0	4.0	5.0
2.0	—	0.25	0.15	0.15	0.10	0.10	0.10	0.10	0.10	0.05	0.05	0.05	0.05	0	0
1.8	—	0.20	0.15	0.10	0.10	0.10	0.05	0.05	0.05	0.05	0.05	0.05	0	0	0
1.6	0.4	0.15	0.10	0.10	0.05	0.05	0.05	0.05	0.05	0.05	0.05	0	0	0	0
1.4	0.6	0.10	0.05	0.05	0.05	0.05	0.05	0.05	0.05	0.05	0	0	0	0	0
1.2	0.8	0.05	0.05	0.05	0	0	0	0	0	0	0	0	0	0	0
1.0	1.0	0	0	0	0	0	0	0	0	0	0	0	0	0	0
0.8	1.2	−0.05	−0.05	−0.05	0	0	0	0	0	0	0	0	0	0	0
0.6	1.4	−0.10	−0.05	−0.05	−0.05	−0.05	−0.05	−0.05	−0.05	−0.05	0	0	0	0	0
0.4	1.6	−0.15	−0.10	−0.10	−0.05	−0.05	−0.05	−0.05	−0.05	−0.05	−0.05	0	0	0	0
—	1.8	−0.20	−0.15	−0.10	−0.10	−0.10	−0.05	−0.05	−0.05	−0.05	−0.05	−0.05	0	0	0
—	2.0	−0.25	−0.15	−0.15	−0.10	−0.10	−0.10	−0.10	−0.10	−0.05	−0.05	−0.05	−0.05	0	0

Note: $\alpha_2 = \dfrac{h_{\text{above}}}{h}$, $\alpha_3 = \dfrac{h_{\text{below}}}{h}$ (Figure 5.4).

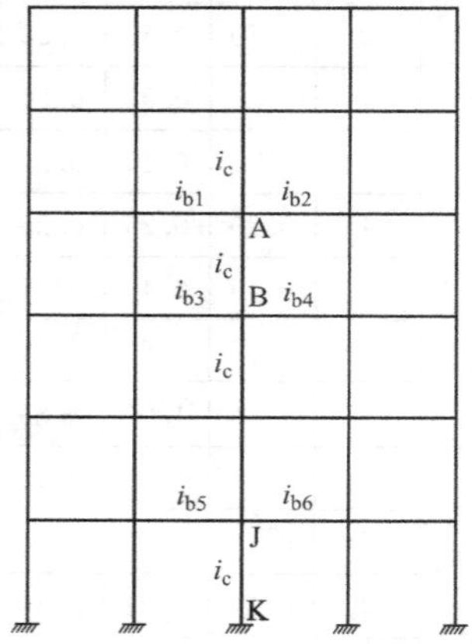

(4) Internal forces of frame

The procedure for calculation of frame internal forces is as follows:

1) Calculating the lateral stiffness of columns by the formula (5-30) and the formula (5-31), respectively.

2) Calculating the shear forces of columns by the following formula:

$$V_{ik} = \frac{D_{ik}}{\sum\limits_{i=1}^{n} D_{ik}} V_i \qquad (5-33)$$

Figure 5.3   The determination of non-bottom column D-value

Where V_{ik} is the seismic shear force of kth column on ith storey, D_{ik} is the lateral stiffness of kth column on ith storey, $\sum\limits_{i=1}^{n} D_{ik}$ is the summation of lateral stiffness of all columns on ith storey, V_i is the seismic shear force of ith storey.

3) Calculating the height h' of column contra-flexure point by the formula (5-32).

4) Calculating the bending moments of column ends by V_{ik} and h'.

5) Calculating the bending moments of beam ends by the bending moment balance of joints between column and beam, furthermore, the shear forces of beams and the axial forces of columns.

(5) Elastic storey drift

The procedure for calculation of elastic storey drift of frame under frequent earthquake is as follows:

1) Calculating the linear stiffness of all columns and beams.

2) Calculating the lateral stiffness of columns D_{ik} and $\sum\limits_{i=1}^{n} D_{ik}$.

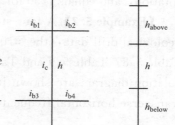

3) Calculating the basic natural vibration period by the formula (5-14) or the formula (5-26).

4) Calculating the horizontal seismic influence coefficient.

5) Calculating the standard value of horizontal seismic action by the base shear method, furthermore, the storey seismic shear force.

Figure 5.4   The adjustment coefficients

6) Calculating the ith storey elastic drift by the following formula:

$$\Delta u_{ei} = \frac{V_i}{\sum\limits_{i=1}^{n} D_{ik}} \tag{5-34}$$

7) Checking the ith storey elastic drift by the following expression:

$$\Delta u_{ei} \leqslant [\theta_e] h_i \tag{5-35}$$

Where h_i is the $i$thstorey height, $[\theta_e]$ is called the limit for elastic storey drift rotation, and equals $1/550$ for reinforced concrete frame and $1/250$ for steel structure, respectively.

(6) Elasto-plastic storey drift

The procedure for calculation of elasto-plastic storey drift of frame under rare earthquake by the simplified method is as follows:

1) Calculating the storey yield shear forces V_y of frame, which are determined by the actual reinforcement and material strength of frame members.

2) Calculating the horizontal seismic influence coefficient, furthermore, the elastic seismic shear forces V_e of each storey.

3) Calculating the elastic storey drift Δu_e by the formula (5-34).

4) Checking the yield strength coefficient of ith storey by the following expression

$$\xi_y(i) = \frac{V_y(i)}{V_e(i)} \tag{5-36}$$

5) Calculating the elasto-plastic drift of storey, which is the first floor for the frame whose ξ_y distributes uniformly along its height or the storey whose ξ_y is the minimum or relatively small (generally $2\sim3$ stories) for the frame whose ξ_y distributes non-uniformly along its height.

$$\Delta u_p = \eta_p \Delta u_e \tag{5-37}$$

Where η_p is called the enhancement coefficient for elasto-plastic storey drift, and could be seen in the *Code for Seismic Design of Buildings* GB 50011—2010.

6) Checking the storey elasto-plastic drift by the following expression

$$\Delta u_p \leqslant [\theta_p] h \tag{5-38}$$

Where h is the height of weak storey, $[\theta_p]$ is called the limit for elasto-plastic storey drift rotation, and equals 1/50 for both reinforced concrete frame and steel structure.

【**Example 5. 2**】 A four-storey teaching building will be built in a seismic region. The geological drill data, the structure information and the seismic information are shown in Table 5.8, Table 5.9 and Table 5.10, respectively. The structural plan, section and calculation diagram are shown in Figure 5.5. Calculating the internal forces of frame under transverse horizontal frequent earthquake and checking the storey elastic drifts.

Geological drill data   Table 5. 8

Soil type	Depth of the soil layer bottom(m)	Shear-wave velocity(m/s)
Mixed backfill soil	1.2	140
Silty clay	8.4	320
Muddy clay	18.2	260
Rock		>500

Structure information   Table 5. 9

Component	Concrete grade	Elastic modulus(MPa)	Section size(mm)
Column	C35	3.15×10^4	500×500
Beam	C30	3.00×10^4	250×600
Slab	C30	3.00×10^4	100

Seismic information   Table 5. 10

Design seismic information	Parameters
Representative value of gravity load	$G_4 = 7970\text{kN}$, $G_3 = 9230\text{kN}$ $G_2 = 9230\text{kN}$, $G_1 = 9860\text{kN}$
Fortification intensity	7
Design basic acceleration of ground motion	$0.15g$
Design seismic group	First
Damping ratio of structure	0.05

Solution:

(1) Determination of the site category.

Thickness of cover layer: $d_{ov} = 18.2\text{m}$

Calculation depth: $d_0 = \min\{18.2\text{m}, 20\text{m}\} = 18.2\text{m}$

Wave propagation time: $t = \dfrac{1.2}{140} + \dfrac{8.4 - 1.2}{320} + \dfrac{18.2 - 8.4}{260} = 0.0756\text{s}$

Equivalent shear-wave velocity: $v_{se} = \dfrac{d_0}{t} = 241\text{m/s}$

Due to $250 \geqslant v_{se} > 150$ and $d_{ov} \in [3, 50]$, the site category is II.

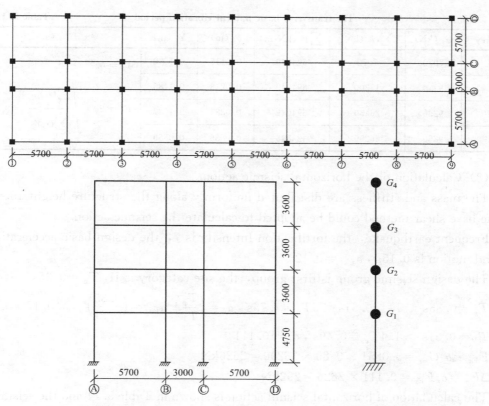

Figure 5.5　The structural plan, section and calculation diagram

(2) Calculation of the transverse basic natural vibration period.

The detailed calculations are shown in Table 5.11, Table 5.12 and Table 5.13, respectively.

The linear stiffness of frame members　　　　　　　　　　　　**Table 5.11**

Member	Location	Linear stiffness EI/l (kN · m)	
		①⑨	②～⑧
Beam	Middle-span	67500	90000
	Side-span	35526	47368
Column	Non-bottom	45573	
	Bottom	34540	

The storey lateral stiffness of frame　　　　　　　　　　　　**Table 5.12**

Storey	Formula	①⑨Column		②～⑧Column		ΣD(kN/m)
		Exterior	Interior	Exterior	Interior	
2～4	$\overline{K}=(i_{b1}+i_{b2}+i_{b3}+i_{b4})/(2i_c)$	$\overline{K}=0.780$	$\overline{K}=2.261$	$\overline{K}=1.039$	$\overline{K}=3.014$	694049
	$\alpha=\overline{K}/(2+\overline{K})$	$\alpha=0.281$	$\alpha=0.531$	$\alpha=0.342$	$\alpha=0.601$	
1	$\overline{K}=(i_{b5}+i_{b6})/i_c$	$\overline{K}=1.029$	$\overline{K}=2.983$	$\overline{K}=1.371$	$\overline{K}=3.977$	423843
	$\alpha=(0.5+\overline{K})/(2+\overline{K})$	$\alpha=0.505$	$\alpha=0.699$	$\alpha=0.555$	$\alpha=0.749$	

The transverse basic natural vibration period   **Table 5. 13**

Storey	G_i (kN)	$\sum G_i$ (kN)	$\sum D$ (kN/m)	δ_i (m)	Y_i (m)	T_1 (s)
4	7970	7970	694049	0.0115	0.1600	$\phi_T = 0.7$
3	9230	17200	694049	0.0248	0.1485	
2	9230	26430	694049	0.0381	0.1237	$T_1 = 2\varphi_T \sqrt{\dfrac{\sum\limits_{i=1}^{n} G_i Y_i^2}{\sum\limits_{i=1}^{n} G_i Y_i}} = 0.51$
1	9860	36290	423843	0.0856	0.0856	

(3) Calculation of the horizontal seismic action.

The mass and stiffness are distributed uniformly along the structure height 15.55m, so the base shear method could be adopted to calculate the seismic action.

Frequent earthquake, the fortification Intensity is 7, the design basic acceleration of ground motion is $0.15g$, $\alpha_{max} = 0.12$.

The design seismic group is first group, the site category is II, $T_g = 0.35s$

$$T_g = 0.35s < T_1 = 0.51s < 5T_g = 1.75s , \ \alpha_1 = \left(\frac{T_g}{T_1}\right)^{\gamma} \alpha_{max} = \left(\frac{0.35}{0.51}\right)^{0.9} \times 0.12 = 0.0851$$

$$T_1 = 0.51s > 1.4T_g = 0.49s , \ \delta_n = 0.111$$
$$F_{Ek} = \alpha_1 G_{eq} = 0.0851 \times 0.85 \times 36290 = 2626kN$$
$$\Delta F_n = \delta_n F_{Ek} = 0.111 \times 2626 = 292kN$$

The calcu lation of horizontal seismil action is shown in Table 5.14 and the seismic action diagram is shown in Figure 5.6.

Calculation of F_i , V_i and Δu_{ei}   **Table 5. 14**

Storey	G_i (kN)	$\sum G_i$ (kN)	H_i (m)	F_i (kN)	V_i (kN)	$\dfrac{h_i}{\Delta u_{ei}}$	$\dfrac{V_i}{\sum G_i}$	
4	7970	7970	15.55	808	1100	2273		0.138	
3	9230	17200	11.95	719	1818	1374	>550 Meeting the requirement	0.106	>0.024 Meeting the requirement
2	9230	26430	8.35	502	2320	1077		0.088	
1	9860	36290	4.75	305	2626	767		0.072	

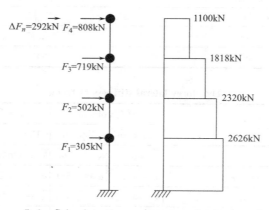

Figure 5.6   Seismic action and storey seismic shear force

(4) Calculation of the internal forces of frame under seismic action.

The detaiced calulntions are shown in Table 5.15，Table 5.16，Table 5.17 and Table 5.18，respectively. The interal force diagrams are shown in Figure 5.7，Figure 5.8 and Figure 5.9，respectively.

Shear force of column and bending moment of column ends for ①⑨ axes　Table 5.15

①⑨Column	Storey	V_i (kN)	$\dfrac{D_{ik}}{\sum D}$	V_{ik} (kN)	y_0	y_1	y_2	y_3	h'(m)	M_{bottom} (kN·m)	M_{top} (kN·m)
Exterior	4	1100	0.0171	18.75	0.350	0	0	0	1.260	23.62	43.87
	3	1818	0.0171	31.00	0.440	0	0	0	1.583	49.08	62.53
	2	2320	0.0171	39.57	0.500	0	0	−0.030	1.693	66.97	75.47
	1	2626	0.0219	57.44	0.647	0	0	0	3.074	176.55	96.27
Interior	4	1100	0.0323	35.47	0.450	0	0	0	1.620	57.46	70.23
	3	1818	0.0323	58.65	0.463	0	0	0	1.667	97.77	113.38
	2	2320	0.0323	74.85	0.500	0	0	0	1.800	134.74	134.74
	1	2626	0.0303	79.54	0.550	0	0	0	2.613	207.80	170.02

Shear force of beam ends and axial force of column for ①⑨ axes　Table 5.16

Storey	AB-span			BC-span			Axial force of column	
	M_{AB}(kN·m)	M_{BA}(kN·m)	V_{AB}(kN)	M_{BC}(kN·m)	M_{CB}(kN·m)	V_{BC}(kN)	Exterior	Interior
4	43.87	24.22	11.95	46.01	46.01	30.67	11.95	18.73
3	86.15	58.91	25.45	111.93	111.93	74.62	37.40	67.90
2	124.56	80.18	35.92	152.33	152.33	101.55	73.32	133.53
1	163.23	105.09	47.07	199.67	199.67	133.11	120.39	219.57

Shear force of column and bending moment of column ends for ②～⑧ axes　Table 5.17

②～⑧Column	Storey	V_i (kN)	$\dfrac{D_{ik}}{\sum D}$	V_{ik} (kN)	y_0	y_1	y_2	y_3	h'(m)	M_{bottom} (kN·m)	M_{top} (kN·m)
Exterior	4	1100	0.0208	22.86	0.402	0	0	0	1.447	33.08	49.22
	3	1818	0.0208	37.80	0.450	0	0	0	1.620	61.24	74.85
	2	2320	0.0208	48.24	0.500	0	0	0	1.800	86.84	86.84
	1	2626	0.0241	63.17	0.613	0	0	0	2.911	183.89	116.16
Interior	4	1100	0.0366	40.19	0.450	0	0	0	1.620	65.10	79.57
	3	1818	0.0366	66.45	0.500	0	0	0	1.800	119.61	119.61
	2	2320	0.0366	84.81	0.500	0	0	0	1.800	152.65	152.65
	1	2626	0.0325	85.24	0.550	0	0	0	2.613	223.69	182.20

Shear force of beam ends and axial force of column for ②～⑧ axes　Table 5.18

Storey	AB-span			BC-span			Axial force of column	
	M_{AB}(kN·m)	M_{BA}(kN·m)	V_{AB}(kN)	M_{BC}(kN·m)	M_{CB}(kN·m)	V_{BC}(kN)	Exterior	Interior
4	49.22	27.44	13.45	52.13	52.13	34.75	13.45	21.30
3	107.93	63.69	30.11	121.02	121.02	80.68	43.56	71.88
2	148.08	93.88	42.45	178.38	178.38	118.92	86.01	148.35
1	203.00	115.47	55.87	219.39	219.39	146.26	141.88	238.73

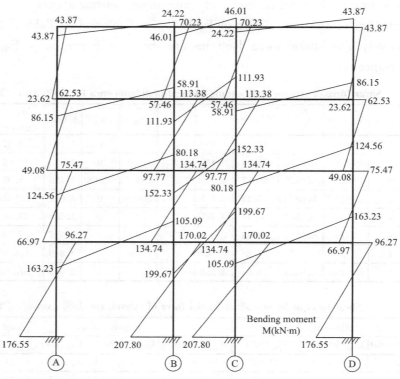

Figure 5. 7   The diagram of bending moment for ①⑨ axes

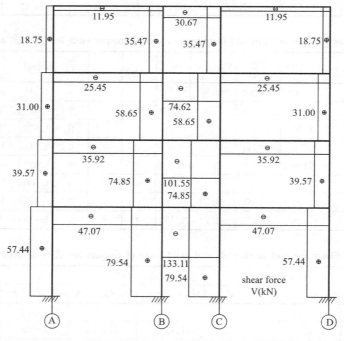

Figure 5. 8   The diagram of shear force for ①⑨ axes

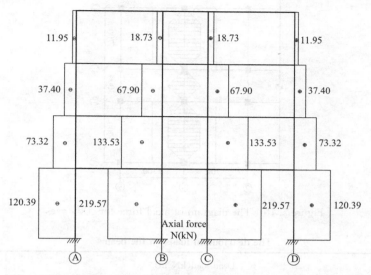

Figure 5. 9　The diagram of axial force for ①⑨ axes

2. Moment distribution method for vertical load

The moment twice-distribution method (distribution-transmission-distribution) is to distribute and transmit the unbalanced bending moments of each node at the same time. The calculation accuracy of the results has met the engineering needs.

【**Example 5. 3**】 Calculating the bending moment of frame in Example 5. 2 under gravity load by the moment twice-distribution method. The loads of roof and floor are shown in Table 5. 19.

Loads in Example 5. 2　　　　　　　　　　　Table 5. 19

Loads	Roof load(kN/m²)	Floor load(kN/m²)		Infill wall(kN/m²)
		Classroom	Corridor	
Dead	5. 65	3. 55	4. 11	2. 92
Live	0. 5	2. 5	3. 5	—

Solution:

(1) Calculation of the distribution loads on the beams of transverse frame.

The loads transmitted from slab to beam are shown in Figure 5. 10. By the equivalent area method, the equivalent rectangular width equal to the area of trapezoidal or triangular loads divided by the beam length, that is, it is 2. 14m for AB-Span and 1. 50m for BC-Span.

The beam weight is $0.25 \times (0.6 - 0.1) \times 25 \times 1.1 = 3.44$kN/m , the wall weight is $2.92 \times (3.6 - 0.6) = 8.76$kN/m . The distribution loads on the beams are shown in Table 5. 20.

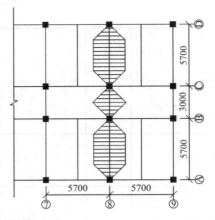

Figure 5.10   The diagram of axial force for ①⑨ axes

The distribution loads on the beams   Table 5.20

Storey		Dead load(kN/m)				Live load(kN/m)
		Slab(kN/m)	Beam(kN/m)	Wall(kN/m)	Total(kN/m)	
AB-Span	4	12.10	3.44	0	15.54	1.07
	3	7.6	3.44	8.76	19.80	5.35
	2	7.6	3.44	8.76	19.80	5.35
	1	7.6	3.44	8.76	19.80	5.35
BC-Span	4	8.50	3.44	0	11.94	0.75
	3	6.17	3.44	0	9.61	5.25
	2	6.17	3.44	0	9.61	5.25
	1	6.17	3.44	0	9.61	5.25

(2) Calculation of the bending moment of beam ends and the distribution factors.

Due to the symmetry of structure, the half of ⑧ frame could be chosen to calculate and is shown in Figure 5.11. The bending moment of beam fined ends, the rotation stiffness of members and the distribution factors are shown in Table 5.21, Table 5.22 and Table 5.23, respectively.

Bending moment of beam fixed ends   Table 5.21

Load type	Storey	Bending moment(kN·m)	
		Side-span	Middle-span
Dead load	4	$15.54 \times 5.7^2/12 = 42.07$	$11.94 \times (3/2)^2/3 = 8.96$
	1~3	$19.80 \times 5.7^2/12 = 53.61$	$9.61 \times (3/2)^2/3 = 7.21$
Live load	4	$1.07 \times 5.7^2/12 = 2.90$	$0.75 \times (3/2)^2/3 = 0.56$
	1~3	$5.35 \times 5.7^2/12 = 14.49$	$5.25 \times (3/2)^2/3 = 3.94$

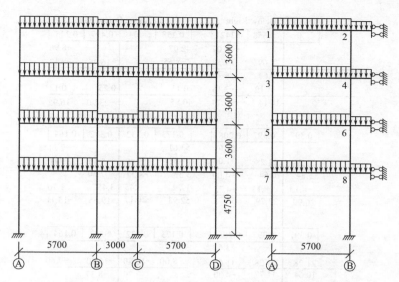

Figure 5. 11　The half of ⑧ frame

Rotation stiffness of members　　　　　　　　　　　　Table 5. 22

Member	Location	Rotation stiffness(kN · m)	Relative stiffness
Beam	Middle-span	$2i_b = 90000$	0. 651
	Side-span	$4i_b = 94736$	0. 686
Column	Non-bottom	$4i_c = 182292$	1. 319
	Bottom	$4i_c = 138160$	1. 000

Distribution factors　　　　　　　　　　　　　　　Table 5. 23

Joint	Sum of relative stiffness	Distribution factors			
		Left	Above	Below	Right
1	$0.686 + 1.319 = 2.005$	—	—	0. 658	0. 342
2	$0.686 + 0.651 + 1.319 = 2.656$	0. 258	—	0. 497	0. 245
3,5	$0.686 + 1.319 + 1.319 = 3.324$	—	0. 397	0. 397	0. 206
4,6	$0.686 + 0.651 + 1.319 + 1.319 = 3.975$	0. 173	0. 332	0. 332	0. 164
7	$0.686 + 1.319 + 1.000 = 3.005$	—	0. 439	0. 333	0. 228
8	$0.686 + 0.651 + 1.319 + 1.000 = 3.656$	0. 188	0. 361	0. 274	0. 178

(3) Distribution and transmission of bending moments.

For example, the distribution and transmission of bending moments of frame under vertical dead load is shown in Figure 5. 12. Firstly, the unbalanced bending moments of each node are distributed at the same time, and then the transmission is done by the transfer coefficient 1/2. Finally, the unbalanced bending moments of each node are redistributed again. Subsequently, the shear force and axial force of frame members will be derived easily by the isolation method.

(4) The diagram of bending moments.

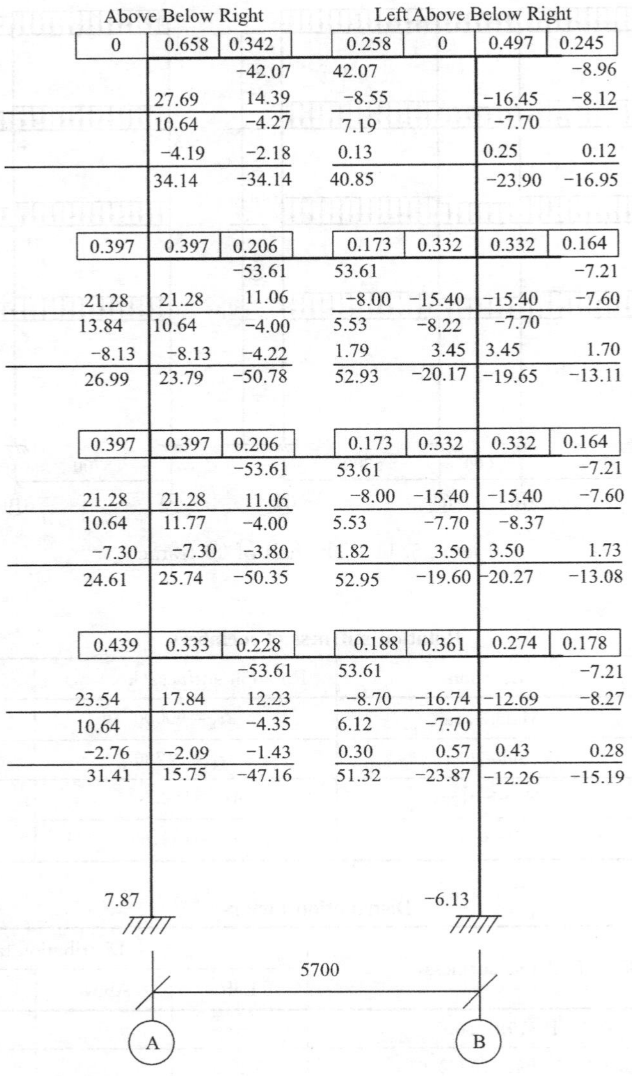

Above Below Right   **Left Above Below Right**

0	0.658	0.342		0.258	0	0.497	0.245
		−42.07		42.07			−8.96
	27.69	14.39		−8.55		−16.45	−8.12
	10.64	−4.27		7.19		−7.70	
	−4.19	−2.18		0.13		0.25	0.12
	34.14	−34.14		40.85		−23.90	−16.95

0.397	0.397	0.206		0.173	0.332	0.332	0.164
		−53.61		53.61			−7.21
21.28	21.28	11.06		−8.00	−15.40	−15.40	−7.60
13.84	10.64	−4.00		5.53	−8.22	−7.70	
−8.13	−8.13	−4.22		1.79	3.45	3.45	1.70
26.99	23.79	−50.78		52.93	−20.17	−19.65	−13.11

0.397	0.397	0.206		0.173	0.332	0.332	0.164
		−53.61		53.61			−7.21
21.28	21.28	11.06		−8.00	−15.40	−15.40	−7.60
10.64	11.77	−4.00		5.53	−7.70	−8.37	
−7.30	−7.30	−3.80		1.82	3.50	3.50	1.73
24.61	25.74	−50.35		52.95	−19.60	−20.27	−13.08

0.439	0.333	0.228		0.188	0.361	0.274	0.178
		−53.61		53.61			−7.21
23.54	17.84	12.23		−8.70	−16.74	−12.69	−8.27
10.64		−4.35		6.12	−7.70		
−2.76	−2.09	−1.43		0.30	0.57	0.43	0.28
31.41	15.75	−47.16		51.32	−23.87	−12.26	−15.19

7.87 (base A)   −6.13 (base B)

5700

A   B

Figure 5.12   Distribution and transmission of bending moments

According to the Figure 5.12, the diagram of bending moment of frame is drawn in Figure 5.13. It is known that the beam ends of frame structure have larger bending moments and more reinforcement, so it is inconvenient to construct. Due to the redistribution of plastic internal forces in statically indeterminate reinforced concrete structures, the bending moments of beam ends could be reduced appropriately and generally multiplied by the reduction factor. According to the engineering experience, the reduction factor is 0.8~0.9 for cast-in-situ reinforced concrete frame and 0.7~0.8 for assembled monolithic reinforced concrete frame. When the bending moments of beam ends decrease, the bending moment of beam mid-span increases. In this way, after adjusting the bending moments of beam ends, the reinforcement at the beam ends can't be reduced only, but the convenient construction can also be achieved.

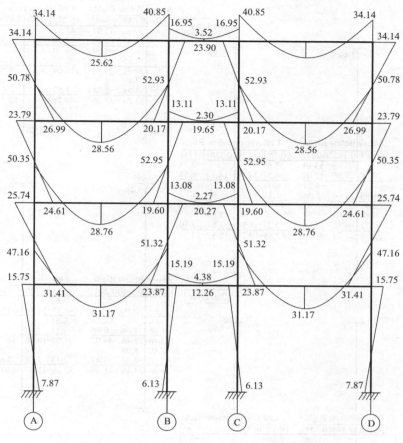

Figure 5.13   The diagram of bending moments

3. Layered method for vertical load

Another method to calculate the internal forces of frame under vertical loads is the so-called layered method, in which it is assumed that the linear stiffness of non-bottom columns should multiply the reduction factor 0.9, and their transmission factor should be replaced by 1/3, instead of 1/2. It is worth mentioning that the more layers of structure, the more obvious advantage of the layered method. As an example, by the layered method, the calculation of bending moment of frame under vertical dead load is shown in Figure 5.14.

5.4.4   Load effect combination and member design

Reinforcement design of members is based on the most unfavorable combination of internal forces at the control section. For frame beams, the beam end section and mid-span section are generally used as control sections, while for frame columns, the top and bottom sections of columns are chosen as control sections. The most unfavorable combination of internal forces is the maximum of internal force combination to control the section reinforcement.

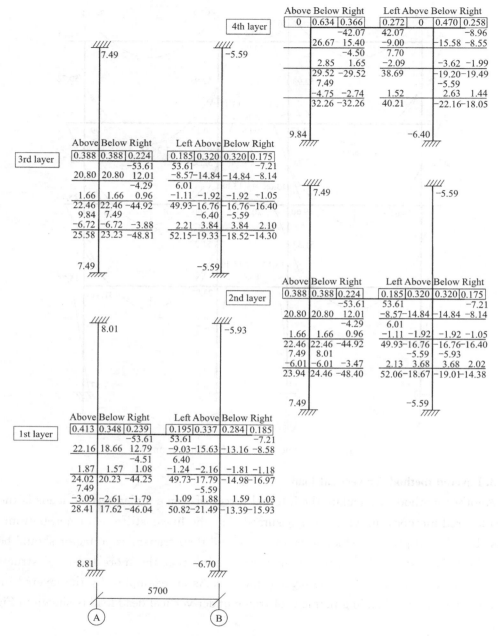

Figure 5.14   Calculation of bending moment of frame by the layered method

1. Load effect combination

Two combinations should be calculated for the ultimate state of bearing capacity.

(1) Fundamental combination

When there is a linear relationship between load and its effect, the design value S_d shall be determined by the most unfavorable value taking from the following combination values

$$S_d = \sum_{i \geqslant 1} \gamma_{G_i} S_{G_{ik}} + \gamma_P S_P + \gamma_{Q_1} \gamma_{L_1} S_{Q_{1k}} + \sum_{j>1} \gamma_{Q_j} \psi_{cj} \gamma_{L_j} S_{Q_{jk}} \tag{5-39}$$

Where γ_{G_i} is the partial safety factor of ith dead load G_{ik}, which is 1.3 when the dead load effect $S_{G_{ik}}$ is unfavorable to structures, and $\leqslant 1.0$ if favorable. γ_P and S_P are the partial safety factor and load effect of pre-stressed load P, respectively. γ_{Q_j} is the partial safety factor of jth live load Q_{jk}, which is 1.5 when the live load effect $S_{Q_{jk}}$ is unfavorable to structures, and 0 if favorable. γ_{L_j} is the adjustment factor of jth live load considering design service life, which is 0.9 for 5 years, 1.0 for 50 years and 1.1 for 100 years, respectively. ψ_{cj} is the combination factor of jth live load, which is 0.7 for civil architecture.

Otherwise, the design value S_d shall be derived as follows:

$$S_d = S\left(\sum_{i \geqslant 1} \gamma_{G_i} G_{ik} + \gamma_P P + \gamma_{Q_1} \gamma_{L_1} Q_{1k} + \sum_{j>1} \gamma_{Q_j} \psi_{cj} \gamma_{L_j} Q_{jk}\right) \tag{5-40}$$

Where $S(\cdot)$ is the effect function of load combination. The symbol $\sum$ and $+$ denote the combination.

(2) Seismic effect combination

$$S_E = \gamma_G S_{GE} + \gamma_{Eh} S_{Ehk} + \gamma_{Ev} S_{Evk} + \psi_w \gamma_w S_{wk} \tag{5-41}$$

Where γ_G is the partial safety factor of gravity load, which is generally 1.2 and should not be greater than 1.0 if the gravity load effect is favorable to the bearing capacity of members. S_{GE} is the effect of the representative value of gravity load. γ_{Eh} and γ_{Ev} are the partial safety factors of the horizontal and vertical seismic actions, respectively, and should be adopted by Table 5.24. S_{Ehk} and S_{Evk} are the effects of standard value of horizontal and vertical seismic action, respectively. ψ_w is the combination value factor of wind load, which is generally 0.0 and 0.2 for such buildings mainly controlled by the wind load. γ_w is the partial safety factor of wind load, which should be 1.4. S_{wk} is the effect of standard value of wind load.

Partial factor of seismic action   Table 5.24

Seismic action	γ_{Eh}	γ_{Ev}
Only horizontal seismic action	1.3	0.0
Only vertical seismic action	0.0	1.3
Mainly horizontal seismic action	1.3	0.5
Mainly vertical seismic action	0.5	1.3

2. Member design

The reinforcement of frame members should be designed by the following expressions

$$\gamma_0 S_d \leqslant R \tag{5-42}$$

Where γ_0 is the importance factor of structure, which should not be less than 1.1, 1.0 and 0.9 for members with safety class I, II and III, respectively. S_d should be adopted by the maximum of the numerical values calculated by the formula (5-39) and the formula (5-40). R is the design value of bearing capacity of structural members.

$$S_E \leqslant \frac{R}{\gamma_{RE}} \tag{5-43}$$

Where S_E should be calculated by the formula (5-41), γ_{RE} is the seismic adjustment coefficient of bearing capacity, which should be selected by Table 5.25 and should be 1.0 for the vertical seismic action calculated only.

Seismic adjustment coefficient of bearing capacity   **Table 5.25**

Material	Structural members	Stress state	γ_{RE}
Steel	Column, beam, brace, gusset plate, bolt and welded joint	Strength	0.75
	Column, brace	Stability	0.80
Masonry	Seismic wall with constructional columns and core columns at both ends	Shear	0.9
	Other seismic walls	Shear	1.0
Concrete	Beam	Bending	0.75
	Column with axial compression ratio less than 0.15	Eccentric compression	0.75
	Column with axial compression ratio no less than 0.15	Eccentric compression	0.80
	Seismic wall	Eccentric compression	0.85
	Each kind of members	Shear, eccentric tension	0.85

Furthermore, the seismic design of members must obey three seismic design principles: ① Strong-column and weak-beam. This principle is to ensure that the plastic hinges appear first in the beams. Because the plastic hinges occur first in columns, structures are prone to collapse due to geometrically variable systems. ② Strong-shear and weak-bend. This principle guarantees that the plastic hinges occur in members without premature shear failure, that is, the shear capacity of members must be greater than the bending capacity of plastic hinges. ③ Strong-joint and strong-anchorage. In order to ensure the ductility requirement of structure, the frame joints and reinforcement anchorage should not be destroyed prematurely before the plastic hinge of the beam can fully play its role.

(1) Seismic design of frame beam

1) Adjustment of design shear force.

According to this principle of strong-shear and weak-bend, the design value V_b of shear force at beam ends should be adjusted by the following expression:

$$V_b = \eta_{vb} \frac{M_b^l + M_b^r}{l_n} + V_{Gb} \tag{5-44}$$

Where η_{vb} is the enhancement coefficient, which is 1.3, 1.2 and 1.1 for seismic grade I, II and III, respectively. M_b^l and M_b^r are the design value of bending moment at the left end and right end of beam, respectively. l_n is the clear span of beam, V_{Gb} is the design value of shear force at the beam ends under the representative value of gravity load, which is analyzed based on the simply-supported beam.

2) Checking of shear-compression ratio.

The shear-compression ratio of a beam is the ratio of the average shear stress in the beam to the design value of concrete compressive strength. Before the diagonal cracks occur in the section of beam, the shear force of member is basically borne by the shear strength of concrete. The tensile stress caused by the shear of stirrup is very low. If the shear-compression ratio of member section is too large, the premature baroclinic failure of concrete will occur. In fact, the limit of the shear-compression ratio of beam is to limit the minimum section of beam.

The design values of shear force at beam ends should also meet the following requirements. When the span-height ratio is not larger than 2.5.

$$V_b \leqslant \frac{0.15 f_c b h_0}{\gamma_{RE}} \tag{5-45}$$

When the span-height ratio is larger than 2.5.

$$V_b \leqslant \frac{0.2 f_c b h_0}{\gamma_{RE}} \tag{5-46}$$

Where f_c is the design value of axial compressive strength of concrete, b and h_0 are the width of beam cross section and the effective height of section, respectively.

3) Checking of shear capacity.

Generally, the seismic shear capacity of frame beam is as follows:

$$V_b \leqslant \frac{1}{\gamma_{RE}} (0.42 f_t b h_0 + 1.25 f_{yv} \frac{A_{sv}}{s} h_0) \tag{5-47}$$

For the frame beam under concentrated load,

$$V_b \leqslant \frac{1}{\gamma_{RE}} (\frac{1.05}{\lambda+1} f_t b h_0 + f_{yv} \frac{A_{sv}}{s} h_0) \tag{5-48}$$

Where λ is the shear-span ratio, which must be between 1.5 and 3.0. f_t is the design value of axial tensile strength of concrete, f_{yv} is the design value of tensile strength of stirrup, A_{sv} is the total area of stirrup cross section, S is the stirrup distance along the axial direction of beam.

(2) Seismic design of frame column

1) Checking of axial compression ratio.

The axial compression ratio $N/(f_c b h)$ refers to the ratio of the design value of combination axial force of column to the product of the total cross-sectional area with the design value of axial compressive strength of concrete. The ductility of column decreases sharply with the increase of axial compression ratio, especially at high axial compression ratio, the effect of stirrups on column deformation is very small. Therefore, the axial compression ratio must be limited to ensure that the column has certain ductility. The axial compression ratio of frame column should not be larger than 0.65, 0.75, 0.85 and 0.90 for seismic grade I, II, III and IV, respectively.

2) Adjustment of design bending moment.

According to this principle of strong-column and weak-beam, at the beam-column joints of seismic grade I, II, III and IV frames, except for the columns with the axial compression ratio less than 0.15 and the top storey. The bending moments at column ends should be adjusted by the following formula:

$$\sum M_c = \eta_c \sum M_b \tag{5-49}$$

Where $\sum M_c$ and $\sum M_b$ are the summation of bending moment of column ends at the beam-column joint and the summation of bending moment of beam ends at the same beam-column joint, respectively. η_c is the enhancement coefficient, which is 1.7, 1.5, 1.3 and 1.2 for seismic grade Ⅰ, Ⅱ, Ⅲ and Ⅳ frames, respectively.

3) Adjustment of design shear force.

According to this principle of strong-shear and weak-bend, the design value V_c of shear force at column ends should be adjusted by the following expression:

$$V_c = \eta_{vc} \frac{M_c^t + M_c^b}{H_n} \tag{5-50}$$

Where η_{vc} is the enhancement coefficient, which is 1.5, 1.3, 1.2 and 1.1 for seismic grade Ⅰ, Ⅱ, Ⅲ and Ⅳ frames, respectively. M_c^t and M_c^b are the design value of bending moment at the top end and bottom end of column, respectively. H_n is the net height of column.

4) Checking of shear-compression ratio.

The design value of shear force at column ends should also meet the following requirements. When the shear-span ratio is not larger than 2.0,

$$V_c \leqslant \frac{0.15 f_c b h_0}{\gamma_{RE}} \tag{5-51}$$

When the shear-span ratio is larger than 2.0,

$$V_c \leqslant \frac{0.2 f_c b h_0}{\gamma_{RE}} \tag{5-52}$$

5) Checking of shear capacity.

Generally, the seismic shear capacity of frame column is as follows:

$$V_c \leqslant \frac{1}{\gamma_{RE}} \left(\frac{1.05}{\lambda + 1} f_t b h_0 + f_{yv} \frac{A_{sv}}{s} h_0 + 0.056 N_c \right) \tag{5-53}$$

Where λ is the shear-span ratio, which must be between 1.0 and 3.0. N_c is the design value of column axial compressive force considering the seismic effect combination, which is not larger than $0.3 f_c A_c$. For the tensile column of frame,

$$V_c \leqslant \frac{1}{\gamma_{RE}} \left(\frac{1.05}{\lambda + 1} f_t b h_0 + f_{yv} \frac{A_{sv}}{s} h_0 - 0.2 N_c \right) \tag{5-54}$$

Where N_c is the design value of column axial tensile force corresponding to V_c. If the numerical value in parentheses should not be less than $f_{yv} h_0 A_{sv}/s$, and $f_{yv} h_0 A_{sv}/s$ should not be less than $0.36 f_t b h_0$.

(3) Seismic design of frame joint

The investigation of earthquake damage shows that the failure of frame joints is mainly caused by the insufficient stirrups at joint core zone. The concrete diagonal cracks at joint core zone, the stirrup yield, even the stirrup breakage and the buckling of reinforcement of columns will occur under the combined action of shear and pressure. Therefore, in order to prevent the shear failure at joint core zone, it is necessary to ensure the strength of concrete at joint core zone and to configure sufficient stirrups.

1) Horizontal shear force at joint core zone.

$$V_j = \frac{\eta_{jb} \sum M_b}{h_{b0} - a_s'} (1 - \frac{h_{b0} - a_s'}{H_c - h_b}) \tag{5-55}$$

Where η_{jb} is the strong-joint coefficient, which is 1.5, 1.35 and 1.2 for seismic grade I, II and III, respectively. h_{b0} and h_b are the effective height of beam section and the height of beam cross section, respectively. a_s' is the distance between the force concurrence joint of beam compressive reinforcement and the compressive edge. H_c is the calculation height of column.

2) Checking of shear-compression ratio.

$$V_j \leqslant \frac{0.3\eta_j f_c b_j h_j}{\gamma_{RE}} \tag{5-56}$$

Where η_j is the confined influence factor of orthogonal beam, which is 1.5 for the case that cast-in-situ construction, the center lines of beam and column superpose, the beam section width of four sides is not less than the half of column section width and the orthogonal beam height is not less than 3/4 of frame beam height, and 1.0 for other cases. b_j and h_j are the width and height of column section at joint core zone, respectively, which are calculated by the *Code for Seismic Design of Buildings* GB 50011—2010.

3) Checking of shear capacity.

$$V_j \leqslant \frac{1}{\gamma_{RE}} (\eta_j f_t b_j h_j + 0.5\eta_j N_c \frac{b_j}{b_c} + f_{yv} A_{svj} \frac{h_{b0} - a_s'}{s}) \tag{5-57}$$

Where N_c is the smaller value of upper-column axial compressive force corresponding to design value of shear force, which is 0.0 if axial tensile force.

5.5 Details of seismic design for frame structures

5.5.1 Stirrups in beam, column and joint

The investigation of earthquake damage and theoretical analysis show that under seismic action, the shear force at the ends of beam and column is the largest, and shear failure is quite prone to occur at these locations. Therefore, the stirrup space should be appropriately densified within a certain range at the ends of beam and column, which is called the stirrup densified area.

1. The stirrup densified area at beam end

The length of the stirrup densified area at beam end, the maximum space of stirrup and the minimum diameter of stirrup should be adopted by Table 5.26, where d is the diameter of longitudinal reinforcement and h_b is the height of beam cross section. The minimum diameter of stirrup should be increased by 2mm when the longitudinal tension reinforcement ratio at beam end exceeds 2%.

Length of stirrup densified area, maximum space and minimum diameter   Table 5. 26

Seismic grade	Length of densified area (mm) (choose the maximum)	Maximum space of stirrups (mm) (choose the minimum)	Minimum diameter of stirrup (mm)
I	$2h_b$,500	$h_b/4$, $6d$,100	10
II	$1.5h_b$,500	$h_b/4$, $8d$,100	8
III	$1.5h_b$,500	$h_b/4$, $8d$,150	8
IV	$1.5h_b$,500	$h_b/4$, $8d$,150	6

The space of stirrup legs in the densified area at beam end should not be larger than the larger value between 200mm and 20 times stirrup diameter for seismic grade I , between 250mm and 20 times stirrup diameter for seismic grade II and III , and 300mm for seismic grade IV .

2. The stirrup densified area at column end

(1) The stirrup densified area

The stirrup densified area of a column should meet the following requirements.

1) At the column end, the maximum value among the cross section height (diameter of circular column), 1/6 of the net height of column and 500mm should be taken.

2) For the lower end of bottom column, the densified area should not be less than 1/3 of the net height of column.

3) The areas of 500mm above and below the rigid ground respectively.

4) The total height of column should be taken as the densified area for the columns with shear-span ratio no more than 2, the columns with the ratio of net height to section height no more than 4 and corner columns of seismic grade I and II frames.

(2) Space and diameter of stirrup at densified area

Generally, the maximum space and minimum diameter of stirrups in the stirrup densified area of column should be adopted by Table 5. 27, where d is the minimum diameter of longitudinal reinforcement of column, and the column root refers to the lower end of bottom column.

Maximum space and minimum diameter of stirrups in the densified area of column   Table 5. 27

Seismic grade	Maximum space of stirrups (mm) (choose the minimum)	Minimum diameter of stirrup (mm)
I	$6d$,100	10
II	$8d$,100	8
III	$8d$,150 (100 at column root)	8
IV	$8d$,150 (100 at column root)	6 (8 at column root)

If the diameter of stirrup for seismic grade I frame column is larger than 12mm and the space of stirrup legs is not larger than 150mm, the diameter of stirrup for seismic grade II frame column is larger than 10mm and the space of stirrup legs is not larger than

200mm, except for the lower end of bottom column, the maximum space of stirrups may be taken 150mm. If the cross section size of seismic grade Ⅲ frame column is not larger than 400mm, the minimum diameter of stirrup may be taken 6mm. If the shear-span ratio of seismic grade Ⅳ frame column is not larger than 2, the diameter of stirrup should not be less than 8mm.

The stirrup space should not be larger than 100mm for the frame columns with the shear-span ratio no more than 2.

(3) Space of stirrup legs in the densified area of column

The space of stirrup legs should not be larger than 200mm for seismic grade Ⅰ, 250mm for seismic grade Ⅱ and Ⅲ, and 300mm for seismic grade Ⅳ.

(4) Volume stirrup ratio in the densified area of column

The volume stirrup ratio ρ_v should meet the following expression:

$$\rho_v \geqslant \lambda_v \frac{f_c}{f_{yv}} \tag{5-58}$$

Where ρ_v should not be less than 0.8% for seismic grade Ⅰ, 0.6% for seismic grade Ⅱ and 0.4% for seismic grade Ⅲ and Ⅳ. f_c is the design value of axial compressive strength of concrete with grade not lower than C35, that is, if the concrete grade is lower than C35, the parameters of C35 will be adopted. f_{yv} is the design value of tensile strength of stirrup. λ_v is the minimum stirrup characteristic value, which should be adopted by Table 5.28.

Maximum stirrup characteristic value in the densified area of column　　Table 5.28

Seismic grade	Stirrup type	Axial compression ratio of column								
		≤0.3	0.4	0.5	0.6	0.7	0.8	0.9	1.0	1.05
Ⅰ	Ordinary stirrup and compound stirrup	0.10	0.11	0.13	0.15	0.17	0.20	0.23	—	—
	Spiral stirrup, compound orcontinuous compound rectangular spiral stirrup	0.08	0.09	0.11	0.13	0.15	0.18	0.21	—	—
Ⅱ	Ordinary stirrup and compound stirrup	0.08	0.09	0.11	0.13	0.15	0.17	0.19	0.22	0.24
	Spiral stirrup, compound orcontinuous compound rectangular spiral stirrup	0.06	0.07	0.09	0.11	0.13	0.15	0.17	0.20	0.22
Ⅲ、Ⅳ	Ordinary stirrup and compound stirrup	0.06	0.07	0.09	0.11	0.13	0.15	0.17	0.20	0.22
	Spiral stirrup, compound orcontinuous compound rectangular spiral stirrup	0.05	0.06	0.07	0.09	0.11	0.13	0.15	0.18	0.20

3. The stirrups in joint core zone

The maximum space and minimum diameter of stirrups in the joint core zone of frame should be in accordance with those requirements of stirrups in the densified area of column. For seismic grade Ⅰ, Ⅱ and Ⅲ frames, the stirrup characteristic value in the joint core zones should not be less than 0.12, 0.10 and 0.08 respectively and the volume stirrup ratio should not be less than 0.6%, 0.5% and 0.4%. In the joint core zone of frame with the column shear-span ratio no more than 2, the volume stirrup ratio should not be less than the larger ones between the upper and lower column ends of core zone.

5.5.2　Anchorage of reinforcement

The seismic anchorage length of longitudinal tensile reinforcement should be calculated by the following formula

$$l_{aE} = \zeta_{aE} l_a \tag{5-59}$$

Where ζ_{aE} is the adjustment factor, which is 1.15 for seismic grade I, II, 1.05 for seismic grade III, 1.0 for seismic grade IV. l_a is the anchorage length of longitudinal tensile reinforcement, which is calculated by Chinese code for design of concrete structures. In addition, the anchorage of longitudinal tensile reinforcement of frame beams and columns in frame joint zone should meet the requirements given by Chinese code for design of concrete structures.

Exercises

5.1　What are the main structural systems of multi-storey and high-rise reinforced concrete buildings? What are the characteristics and applicable scope of them?

5.2　How to determine the seismic grade of multi-story and high-rise reinforced concrete buildings?

5.3　How to calculate the natural vibration period of the frame structure? How to determine the horizontal seismic action of frame structure?

5.4　How to calculate the internal forces of frame structure under the horizontal seismic action and the vertical loading, respectively?

5.5　Why must be the axial compression ratio of frame columns limited?

5.6　How to understand the basic principles of "strong-column and weak-beam", "strong-shear and weak-bending", "strong-joint and weak-component" in the seismic design of frame structure? How to achieve them?

5.7　The horizontal forces at the floor-level of the three-story frame structure and the relative value of the stiffness-per-length of each member are shown in the Figure 5.15. Calculating the internal forces by the D-value method and drawing the internal force diagram of structure.

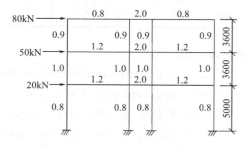

Figure 5.15　The frame structure under the horizontal forces

Chapter 6　Seismic Resistance of Underground Structures

6.1　Introduction

After the 1950s and 1960s, with the development of economic construction in various countries, the process of urbanization accelerated, the development of underground structure space gradually gained attention, and the construction of underground structures gradually increased, such as subways, underground parking lots, underground streets, tunnels, underground pipelines, etc. It is generally believed that the seismic performance of underground structures is better than that of ground structures. However, the seismic response of underground structures and the corresponding seismic design methods are quite different from those of ground structures. After several major earthquakes, the earthquake damage of underground structures is also very serious, and the consequences are catastrophic. In the earthquake in the southern part of Hyogo Prefecture in Japan in 1995, the underground drainage pipes and natural gas pipes were not seriously damaged only, but also many subway stations and section tunnels were damaged. Especially in the Hanshin earthquake that occurred in Japan on January 17, 1995, the underground structures of Kobe City, such as underground stations and underground shopping malls, suffered various degrees of damage, especially the Daikai subway stations and the section tunnels were quite serious damage, as shown in Figure 6.1. In the earthquake of Taiwan on September 21, in 1999, the underground water pipeline of Fengyuan Water Supply Plant in Taizhong County was distorted (steel pipe diameter is 2m, wall thickness is 18mm, Figure 6.2). The M8.0 earthquake that occurred in Wenchuan in 2008 caused the collapse of several tunnels (Figure 6.3 and Figure 6.4). This has caused researchers to pay attention to the seismic design of underground structures.

The underground structure is different from the ground structure. The analysis and study of the safety and stability of underground structure are not only related to the structure itself, but also closely related to the surrounding environment, engineering geological conditions and construction process, that is, the stability of the surrounding rock greatly affects the safety of the underground structure. The development of underground structure calculation theory has been more than 100 years, but the underground structure has long been in the stage of "experience design" and "experience construction".

This chapter consists of 4 sections, mainly including the earthquake damage of under-

ground structures, the earthquake resistance of underground structures, and the calculation method of seismic design of underground structures. In this chapter, we need to master the seismic design theory and method of underground structures and understand the sci-
entific nature of seismic design of underground structures to ensure the seismic safety of
large underground projects and underground lifeline projects and minimize the loss of life
and the economic damage caused by earthquakes.

Figure 6.1   Structural damage of Daikai station

Figure 6.2   Pipe distortion

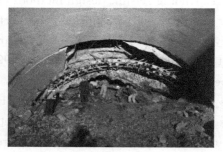

Figure 6.3   Earthquake collapse
at the exit of Longxi Tunnel

Figure 6.4   Slope collapse and collapse
at the exit of Longdongzi Tunnel

6.2   Earthquake damage of underground structures

6.2.1   Introduction

Underground structures are widely used in energy, transportation, communications,
urban construction national defense engineering and other aspects. With the development of
industrial and agricultural production and the increasing degree of urbanization, the importance of underground structures is becoming more and more obvious. At present, underground space has been developed as a resource. Most of China are in the earthquake fortification areas, with the expansion of underground structure construction, the seismic design and safety evaluation of underground structures will become an important issue for engineering designers. Many large earthquakes in history have made underground structures

suffer a certain degree of damage. For example, in the 1995 Hanshin earthquake, some subway stations and section tunnels in Kobe City suffered various degrees of damage. Among them, the Daikai station is most serious, and more than half of the middle columns are completely collapsed, so that the roof collapsed and the overburden layer settled a lot and the maximum settlement is 2.5m. During the earthquake in San Francisco in 1906, three major water pipelines were destroyed, thousands of ruptures occurred in urban water distribution networks, the fire water source was cut off, so that the fire caused by the earthquake could not be extinguished in time. The fire burned for three days and two nights, causing 800 people were killed and property losses were $400 million. In February 1971, in the M6.6 earthquake in San Fernando City, USA, the gas pipes and water pipes were severely damaged. In the M7.8 Tangshan earthquake in 1976, the water supply system in Tangshan City was completely paralyzed, and the water supply was basically restored after one month of repair.

The existing seismic design codes for underground structures are generally very simple, and it is difficult to adapt the development of underground structure construction in the earthquake zone, which makes the study of seismic design of underground structures very necessary. The earthquake damage investigation is to summarize the types and influencing factors of the earthquake damage of the underground structure, and toanalyze the mechanism of structural damage, and to provide inspiration for establishing reasonable analytical models and design methods.

6.2.2　Tunnel earthquake damage

The tunnel extends widely, the site conditions along thesubway line are complex and changeable, and the site soil types are different. The difference of earthquake damage types of tunnel is closely related to geological conditions, focal distance, seismic intensity, seismic wave characteristics, direction of seismic action, relative stiffness of tunnel and surrounding rock, lining construction conditions, construction methods, construction difficulty and whether collapses in the construction process or not. Tunnels are linear structures. Under the action of seismic loads, the main types of earthquake damage are as follows:

(1) Shear displacement of lining. When a tunnel passes through an area where geological conditions vary greatly, such as a tunnel crossing a fault or a transition zone of soil from hard to soft, the relative displacement of surrounding rock causes the tunnel to be separated vertically or horizontally, or at the same time in both directions, resulting in serious shear failure of the whole tunnel. The shear failure of a highway tunnel during the Wenchuan earthquake is shown in Figure 6.5.

(2) Lining cracks. When the tunnel passes through the liquefaction zone of sand or where the shape and stiffness of the structure section change obviously, such as the entrance and exit of the tunnel and the intersection of the tunnel and the station, the lining

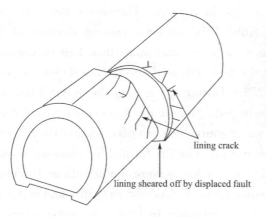

Figure 6.5 Lining shear failure diagram

Figure 6.6 Taoguan tunnel cracking

cracking is prone to occur. Figure 6.6 shows that the circular-arc end wall of Taoguan Tunnel on Dufen Second-Class Road is severely cracked and the end wall and arch ring are loosened. The crack is wide, and there is no connecting steel bar between the end wall and the lining.

6.2.3 Earthquake damage at the subway station

This section mainly describes the earthquake damage occurred in the 1995 Osaka-Kobe earthquake in the subway in Kobe City, and provides some opinions and analysis on the cause of damage for the reference of subway seismic design research.

The railway facilities in Kobe City mainly included JR, Hanshin, Shanyang, Kobe electric railway, Kobe high-speed railway, Municipal subway and Northern god express railway. Among them, there were five routes across the urban area: Kobe, Shanyang, Kobe electric railway, Kobe high-speed railway and City camp railway (Table 6.1). The total length of underground part of the line was about 21.4 km, and the total number of stations was 21.

The line length and station numbers for the underground part of the Kobe subway   Table 6. 1

Name	Construction time	Approximate length of the line (km)	Number of stations
Hanshin electric railway	1931~1936	3. 4	3
Kobe electric railway	1962~1967	0. 4	1
Kobe high speed railway	1962~1967	6. 6	6
Municipal subway	1972~1985	9. 5	9
Shanyang electric railway	1982~1997	1. 5	2
Total		21. 4	21

The interior structure of station consists of a platform that doubles as a parking lot, a central hall with ticketing facilities, a fan room, and an electric room. The cross section of structure varies greatly with the number of layers and the number of spans. In addition, there are transfer sections for the uplink and downlink lines before and after the station, and the structure is a large-span underground structure without the middle columns.

Among the six subway stations of the Kobe high-speed railway, theDaikai station and the Nagata station were seriously affected, and other stations were little affected, and only the concrete structures were cracked. This section will show the disaster situation at the Daikai station on the Kobe high-speed railway in detail.

The Daikai station was built in 1962 and was constructed using the open cut method. The station is 120m long and uses a side platform. There are the standard section for the platform part and the central hall section for the ticket gate, which are shown in Figure 6. 7 (a) and Figure 6. 7 (b), respectively. In the central hall section, the roof, floor, side wall and center column are cast-in-place reinforced concrete structures. The spacing between the middle columns is 3. 5m.

The original structure was designed without considering seismic factors. However, the design is very conservative, the safety factor is very high, and the safety factor of middle column is up to 3, that is, the structure was destroyed under three times normal load. During the earthquake, over 30 of 35 middle columns of the station suffered serious damage, and the total length was about 110m, which collapsed completely. The top plate of the structure was folded at 1. 75~2. 00m away from the middle column, the overall section shape becomes M-shape, and the maximum settlement of surface is about 2. 5m. Therefore, Daikai station was severely damaged by the earthquake and was completely unusable, which has been paid attention to. The vertical schematic diagram of Daikai station damage is shown in Figure 6. 8. According to the damage situation, the station could be divided into Section A, Section B and Section C.

The Section A is a standard structure on the side of Nagata station, the damage is most serious, and most of the middle columns are almost completely crushed. Because the rigid joints at both ends of the top plate and the outer side of the upper part of the upper side of the side wall is damaged by bending after the collapse of the middle pillar, the up-

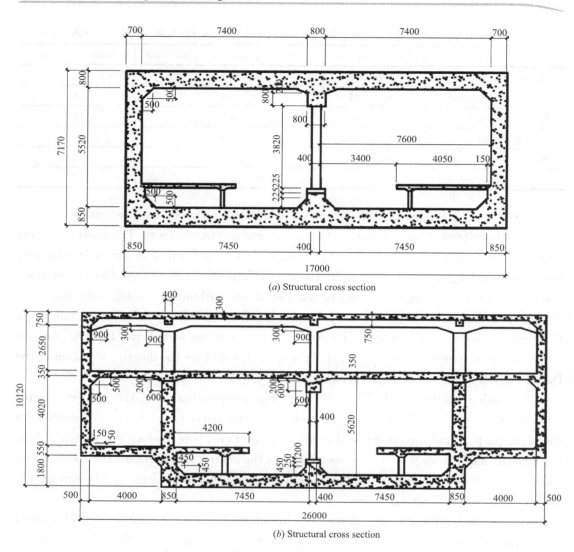

(a) Structural cross section

(b) Structural cross section

Figure 6.7   Typical section of the Daikai station

per top plate is 1.75~2.00m away from the left and right sides of the middle column (main steel bar bending position) is bent. The largest amount of collapse is the slightly westward position of the center of the roof, and the overall section shape becomes M-shaped [Figure 6.9 (a), Figure 6.10]. The collapse of the roof caused a surface main road parallel to it to collapse in the range of 90m long, with a maximum of 2.5 m (Figure 6.11). Within 2m distance from the center line of the top plate, the longitudinal crack width is 150~250mm. Some of the damaged pillars retain some cement, and quite a few of them have broken and fallen off. Some of the 9mm stirrups with a spacing of 35cm are detached together, while others are bent (Figure 6.12). After the pillars are damaged at the upper, lower or both ends, they are shaped like crushed lanterns, the axial reinforcement is bucked in a roughly symmetrical shape to the left or right, or is bent to the left or right. (Figure 6.12 and Figure 6.13). The concrete at the upper end of the side wall is peeled off,

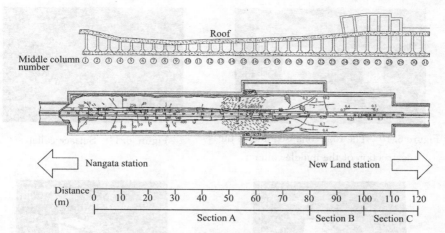

Figure 6. 8　Vertical schematic diagram of Daikai station damage
Note: The figures in the figure indicate the crack width. (Unit: mm)

and in some positions, the main reinforcing bar on the inner side of the side wall is bent, so that the side wall is slightly inwardly bulged, and obvious water leakage can be seen.

The Section B is a two-layer structure [Figure 6. 9 (b)], and the damage is the lightest. Of the six center pillars on the second floor of the basement, two near the Section A and one near the Section C were damaged, and the remaining three were only slightly damaged. Since the covering of this part is only 1. 9m, and the structural safety factor is large, the damage is unexpected.

The structural form of the Section C is similar to that of the Section A, but the degree of damage is lighter than the Section A. In the Section C, shear damage occurs in the lower part of the middle column and the axial reinforcement are bucked [Figure 6. 9 (c)], so that the upper roof is sunken by about 5 cm. In this area, there are no cracks or concrete detachment on the side walls.

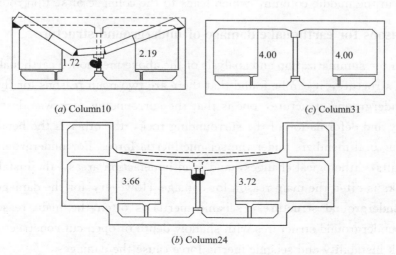

Figure 6. 9　Section damage

149

Figure 6. 10   The roof collapsed due to the
destruction of the middle column

Figure 6. 11   Surface collapse

Figure 6. 12   The damage of the middle
column of Daikai station No. 20

Figure 6. 13   The damage of the middle
column of Daikai station No. 20

On the whole, the Daikai station is a slender box-shaped structure. Under the action of ground motion, the intersection of the roof and the side wall of the station structure are firstly bent and a plastic hinge formed, most of the cover soil weight on the top plate is transferred to the center column. Under the combined action of the soil gravity from the roof and the dynamic stress in the middle column caused by ground motion, bending damage occurred in the middle column, which leads to the collapse of station roof.

6. 2. 4   Reasons for earthquake damage of underground structure

Through the summarization and analysis of the above-mentioned earthquake damage of underground structure, it can be found that there are two main reasons for the earthquake damage of underground structure, one is that the surrounding rock was destroyed due to the instability and deformation of the surrounding rock, the other is the bending or shear failure of structural members under the seismic inertia forces. For underground structures located in faults, lithological changes, and sand liquefaction areas, the instability of surrounding rocks is often the main reason for damage. However, for the damage near the openings of underground structures, seismic inertia is often the main reason for damage. For the underground structures with shallow depth or open-cut construction, the surrounding rock instability and seismic inertia force cause the damages.

In addition, the geological conditions, the stratigraphic sequence of low coastal zone, the vertical and horizontal vibrations, the thickness of overburden layer and the structural form are also have an effect on the underground structure.

6.3   Summary for seismic resistance of underground structures

6.3.1   Seismic response characteristics of underground structures

For ground structures and underground structures, under seismic ground motions, although the natural vibration characteristics of structure and the ground vibration field have a great influence on the dynamic response of structure, for ground structures, the change in shape, mass, and stiffness of the structure, that is, the change of the self-vibration characteristics, has a great influence on the structural response and can cause qualitative changes. However, for underground structures, the main factor that plays a role in the response analysis is the characteristics of foundation motion. Generally, the effect of structure shape change on the structural response is relatively small, with only a change in the amount produced. Therefore, in the current research, the research on the self-revitalization of ground structure accounts for a large proportion, while for the underground structure, the research on the vibration of ground base accounts for a relatively large proportion.

6.3.2   Dynamic characteristics of soil under earthquake

For the earthquake resistance of ground structures, the importance of site foundation is self-evident. Because the ground structures are built on the foundation, the earthquake can't keep the foundation stable (such as sand liquefaction, landslide, etc.), which directly endangers the safety of ground structure. At the same time, the vibration characteristics of ground structure also depend on the transmission of seismic energy by the foundation. Therefore, the earthquake damage of ground structure, in addition to the structure itself, can always find an explanation from the ground.

The underground structure is completely surrounded by the surrounding soil. On the one hand, the deformation of underground structure is affected by the displacement of surrounding soil; on the other hand, the soil plays a role in restraining and restricting the deformation of underground structure. The deformation of underground structure during the earthquake is carried out under the complicated and delicate conditions. Therefore, when it comes to the earthquake resistance of underground structure, the role of site has to be talked.

When selecting the underground structure site, we should master the relevant information of seismic activity, engineering geology and seismic geology according to the project needs, and make comprehensive evaluations for the favorable, unfavorable and dangerous sections under the earthquake. In the case of unfavorable locations, avoidance requirements should be made; effective measures should be taken when they cannot be avoided. For the dangerous areas, it is strictly forbidden to construct underground structures of Class A and Class B, and underground structures of Class C should not be constructed.

6. 3. 3　Dynamic interaction of soil-structure system

When the underground structure and the ground structure are connected as a whole, the seismic waves propagating through the site soil from the source make the structural system to vibrate. At the same time, the inertial force generated by the structural system acts as a new source and in turn acts on the site, which causes additional ground motion to act on the structural system. This phenomenon is called the dynamic interaction of the soil-structure system.

The dynamic interaction of soil-structure system generally includes the interaction caused by the inertial force and a so-called constrained ground motion interaction, which is also called motion interaction (kinematic interaction).

6. 3. 4　Classification of seismic analysis methods for underground structures

The difference between the characteristics of dynamic response of underground structures and that of ground structures determines the different methods of seismic analysis. However, before the 1970s, the seismic design of underground structure basically followed the seismic design of ground structure. After the 1970s, the seismic design of underground structure gradually formed independent theory system.

The method for the seismic analysis of underground structure includes the static method (a simplified method for early seismic design of ground structures), the seismic coefficient method (also known as the pseudo-static method, the inertial force method) and the response displacement method, which are practical methods used in the seismic design of general underground structures. However, for important underground structures (such as underground railways) buried in weak strata, dynamic analysis of seismic response and dynamic model test are often performed.

Theseismic analysis methods of underground structures can be divided into three types: prototype observation, model experiment and theoretical analysis. The theoretical analysis methods can be further subdivided and shown in Figure 6. 14.

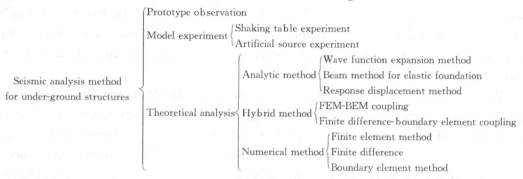

Figure 6. 14　Classification of seismic analysis methods for underground structures

6. 4   Seismic calculation methods of underground structures

The calculation methods for seismic design of underground structures are developed with the continuous development of the understanding of the dynamic response characteristics of underground structures, as well as the investigation, analysis and summary of the earthquake damage of underground structures in recent earthquakes and the deep related research. Before the middle of the 20th century, underground space has not yet been developed on a large scale, and the construction of underground structures has not been greatly developed. Both the size of the unit and the total number are at a lower level. After the middle of the 20th century, with the acceleration of the construction process of various countries, the urbanization process accelerated, and underground engineering has become a valuable resource for human development. Various types of underground structures face extensive development, and earthquake resistance has become a problem that has to be considered in the seismic calculation of underground structure. Based on the calculation methods for seismic design of ground structure and the further study on the response characteristics of underground structures under the seismic ground motion, the pseudo-static design method, the response spectrum theory and the response displacement method become the main methods for seismic design of underground structures during this period.

6. 4. 1   Pseudo-static method

With the continuous development of the earthquake resistance theory of ground structures, the seismic theory of underground structures has gradually developed accordingly. In the 1950s, the seismic design theory of underground structures was based on the static method, which analyzed the seismic forces of underground structures. In the late 1960s, during the construction of BART, the rapid metro transport system in San Francisco Bay, the United States, a new design idea was proposed, such as the ductility of the underground tunnel structure to absorb the forced deformation without losing its static load capacity, and based on this theory, the seismic design criteria was proposed.

The pseudo-static method is also called the static method and the inertial force method, including the seismic coefficient method, the equivalent seismic load method and the equivalent static load method. The main idea of this method is to convert the dynamic load during the earthquake into a static load, and then calculate the response of the structure according to the static force.

(1) Basic assumption

The underground structure could be regarded as an absolute rigid body and has the same seismic response as the surrounding stratum or rock. The maximum inertial force generated by the structure under the seismic ground motion is regarded as the seismic action, and then the internal force and deformation of structure are calculated by combining the influence of the inertial force of the overlying soil, the earth pressure of the ground motion and the internal dynamic water pressure.

(2) Application scope

The method issuitable for the structures, in which the inertia forces are the main factor. For underground structures that are much larger or stiffer than the surrounding strata, the seismic response is consistent with the surrounding rock without relative displacement, and the calculation method provides a reference.

6.4.2 Design response spectrum method

After the San Babara earthquake in 1925, the concept of seismic coefficient was proposed in the United States. After the Long Beach earthquake in 1933, the lateral force coefficient of seismic design was promulgated. It was revised and supplemented in 1937 to gradually consider the dynamic characteristics, but it is still a static model. In 1943, M. A. Biot proposed the concept of response spectrum. In 1948, G. W. Honser proposed the dynamic method of seismic calculation based on response spectrum theory. After the first World Seismic Engineering Conference in 1958, the response spectrum theory was basically adopted in the seismic codes of various countries. The theory of seismic response spectrum establishes the relationship between seismic characteristics and structural dynamic characteristics, but it maintains the original static theory form.

The design response spectrum method is generally applied to the separation of semi-buried structures or underground structures and surrounding rocks. It is considered that the structure displacement at the earthquake duration equals at most the displacement of the surrounding formation, and the "displacement ratio" of soil to the structure depends on the stiffness of the soil to the structure. Therefore, if the displacement distribution and the displacement transmission ratio in the formation are known at the earthquake duration, the structural deformation shape in the earthquake can be obtained, and the seismic design of structure can be done.

6.4.3 Time history analysis method

In this method, the structure is considered as an elasto-plastic vibration system to establish a two-dimensional or three-dimensional soil-structure model directly subject to the seismic waves, and solved by the stepwise integration method. The method considers the three elements of ground motion (amplitude, spectrum and duration) and also considers the dynamic characteristics of structure. The specific and clear requirements are proposed in seismic wave input, structural model design, mass matrix and damping matrix, restoring force model determination and calculation methods, and the reliable design of the structure can be obtained, so it is a more advanced method in the dynamic calculation. For important underground structures, such as underground railway stations, control centers, deep-missed missile silos and other major projects, the dynamic analysis of underground structures can be carried out by plane finite element method or three-dimensional finite element method.

6. 4. 4　**Response displacement method**

Before the 1960s, seismic design methods for underground structures continued to use ground seismic design methods. With the deep understanding of the seismic response analysis of underground structures, it is found that the response laws of underground structures and ground structures in earthquakes are very different. Since the apparent density of underground structure is much smaller than that of the surrounding soil layer, the inertial force of structure is also small. The underground structure is constrained by the surrounding soil, and the vibration energy greatly attenuates. Therefore, it is considered that the seismic characteristics (acceleration, velocity and displacement) of underground structure are basically consistent with the foundation.

In the early 1970s, Japanese scholars firstly proposed the response displacement method based on the above ideas, which was used for the longitudinal seismic design of underground structures such as underground pipelines and tunnels. In the late 1970s, the response displacement method was gradually used for the seismic design of the cross-section of underground structures, but the seismic action only considered the formation displacement. With the deep research on the response displacement method, this method has been applied in many codes in China, and the soil deformation, shear around the structure and the structural inertia are considered in the calculation process.

In the response displacement method, the structure model is built by the beam-spring elements, where the beam elements simulate the underground structure and the spring elements (forward and tangential) simulate the interaction between the structure and the soil. Firstly, calculating the maximum difference of the soil displacement at the top and bottom of structure, and then applying the displacement difference on the spring end away from the structure and applying the structural inertial force. Shear stress is applied directly on the beam elements, and the structural response is calculated using a static method. The detailed steps are as follows:

(1) Considering the soil deformation, the soil shear stress and the inertia force of structure, the surrounding soil can be used as the foundation spring to support structure, and the structure can be modeled by the beam elements, as shown in Figure 6. 15.

where　k_v——structural top plate tensioning foundation spring stiffness (N/m);

　　k_{sv}——structural top floor shear foundation spring stiffness (N/m);

　　k_h——structural sidewall compression spring stiffness (N/m);

　　k_{sh}——structural sidewall shear foundation spring stiffness (N/m);

　　τ_U——shear force acting on the unit area of the structural roof (Pa);

　　τ_B——shear force acting on the unit area of the structural floor (Pa);

　　τ_s——shear force acting on the unit area of the structural side wall (Pa);

　　k_n——circular structure sidewall compression foundation spring stiffness (N/m);

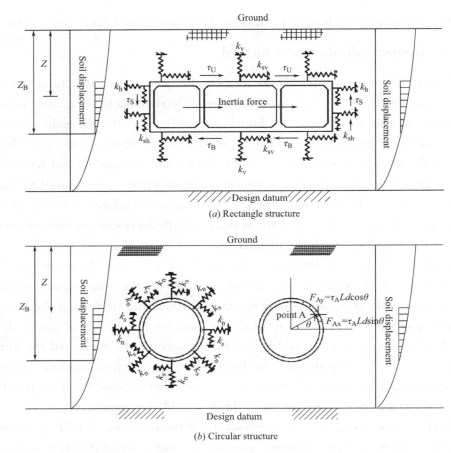

Figure 6.15   Response displacement method for calculation of lateral seismic response

k_s——circular structure sidewall shear foundation spring stiffness (N/m);

τ_A——shear stress at point A (Pa);

F_{AX}——the horizontal nodal force acting on the point A (N);

F_{AY}——the vertical nodal force acting on the point A (N);

θ——the angle between the normal and horizontal directions at the point A of the interface between soil and structure (Pa).

(2) The foundation spring stiffness can be calculated according to the formula (6-1), or it can be calculated according to the static finite element method.

$$k = KLd \tag{6-1}$$

where k——compression, shearing foundation spring stiffness (N/m);

K——the bed coefficient (N/m³) can be obtained according to the current national standard 《Code for Geotechnical Investigation of Urban Rail Transit》 GB 50307—2012;

L——concentrated spring spacing of foundation (m);

d——calculated length of the formation along the longitudinal direction of the underground structure (m).

(3) When using theresponse displacement method to calculate the seismic response of underground structures, the relative displacement of the stratum, the inertial force of the structure and the shear force around the structure should be considered (Figure 6. 16). For the case of the uniform distribution of the stratum, no sudden change in the shape of the section of the structure, and no seismic safety evaluation of the engineering site, the stratum displacement is calculated by the formula (6-2).

$$U(z) = \frac{1}{2} u_{max} \cos \frac{\pi z}{2H} \tag{6-2}$$

where　$U(z)$ ——horizontal displacement of the stratum at depth z during an earthquake (m);

　　　z——depth (m);

　　u_{max} ——the maximum displacement (m) of the site surface, according to Table 6. 2 and Table 6. 3;

　　H——distance from ground to seismic reference plane (m).

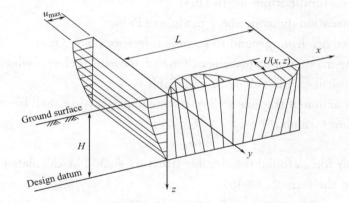

Figure 6. 16　Formation displacement along the depth and axial direction of the tunnel

The relative displacement of the stratum should be calculated according to the formula (6-3).

$$U'(z) = U(z) - U(z_B) \tag{6-3}$$

where　$U'(z)$ ——relative displacement of the free formation relative to the bottom of the structure at depth z (m);

　　$U(z)$ ——seismic response displacement of free formation at depth z;

　　$U(z_B)$ ——seismic response displacement of free formation at depth z_B of the bottom of the structure.

The effect of the relative displacement of the formation can be achieved by applying a forced displacement in the horizontal direction of the joint of the non-structural joint of the foundation spring in the model, calculated according to the formula (6-4):

$$P(z) = kU'(z) \tag{6-4}$$

(4) The structural inertia force should be calculated as follows:

$$f_i = m_i \ddot{u}_i \tag{6-5}$$

where f_i——inertial force acting on the structure i unit (N);

   m_i——structure i unit quality (kg);

   $\ddot{u}_i$——acceleration of structure i unit, taking the peak acceleration (m/s^2), according to Table 6.4 and Table 6.5.

(5) The shear stress of the top and bottom plates of rectangular structure should be calculated according to the following formulas:

$$\tau_U = \frac{\pi G}{4H} u_{max} \sin \frac{\pi z_U}{2H} \tag{6-6}$$

$$\tau_B = \frac{\pi G}{4H} u_{max} \sin \frac{\pi z_B}{2H} \tag{6-7}$$

where τ_U——structural roof shear (Pa);

   τ_B——structural floor shear (Pa);

   z_U——structural roof buried depth (m);

   z_B——structural bottom depth (m);

   G——formation dynamic shear modulus (Pa);

   H——distance from ground to seismic reference plane (m);

   u_{max}——the maximum displacement (m) of the site surface, which shall be in accordance with Table 6.2 and Table 6.3.

The shearing action of the side wall of rectangular structure shall be calculated according to the following formula:

$$\tau_s = (\tau_U + \tau_B)/2 \tag{6-8}$$

(6) The shear force around the circular structure should be calculated according to the formula (6-9) and the formula (6-10).

$$F_{AX} = \tau_A L d \sin\theta \tag{6-9}$$

$$F_{AY} = \tau_A L d \cos\theta \tag{6-10}$$

Horizontal displacement of surface level of Class II sites u (m)　　Table 6.2

Ground motion peak acceleration partition	0.05g	0.10g (0.15g)	0.20g (0.30g)	0.40g
Fortification intensity	6	7	8	9
More earthquakes	0.02	0.04(0.05)	0.07(0.10)	0.14
Design earthquake	0.03	0.07(0.10)	0.13(0.20)	0.27
Rare earthquake	0.08	0.15(0.21)	0.27(0.35)	0.41

Peak displacement adjustment factor Ψ_u　　Table 6.3

Venue category	Peak acceleration of ground motion in Class II sites (g)			
	0.05	0.10 (0.15)	0.20 (0.30)	$\geqslant$0.40
I	0.75	0.80 (0.85)	0.90 (1.00)	1.00
II	1.00	1.00 (1.00)	1.00 (1.00)	1.00
III	1.20	1.25 (1.40)	1.40 (1.40)	1.40
IV	1.45	1.55 (1.70)	1.70 (1.70)	1.70

Horizontal acceleration of surface level of Class II sites a_{max}   Table 6.4

Ground motion peak acceleration partition	$0.05g$	$0.10g(0.15g)$	$0.20g$ $(0.30g)$	$0.40g$
Fortification intensity	6	7	8	9
More earthquakes	$0.02g$	$0.04g(0.05g)$	$0.07g(0.10g)$	$0.14g$
Basic earthquake	$0.05g$	$0.10g(0.15g)$	$0.20g(0.30g)$	$0.40g$
Rare earthquake	$0.11g$	$0.21g(0.32g)$	$0.38g(0.57g)$	$0.64g$
Extremely rare earthquake	$0.15g$	$0.30g(0.45g)$	$0.58g(0.87g)$	$1.08g$

Note: The values in parentheses are used to design areas with basic seismic accelerations of $0.15g$ and $0.30g$, respectively.

Peak acceleration adjustment factor ψ_a   Table 6.5

Venue category	Peak acceleration of ground motion in Class II sites a (g)			
	0.05	0.10 (0.15)	0.20 (0.30)	$\geqslant 0.40$
I	0.80	0.82 (0.83)	0.85 (0.95)	1.00
II	1.00	1.00 (1.00)	1.00 (1.00)	1.00
III	1.30	1.25 (1.15)	1.00 (1.00)	1.00
IV	1.25	1.20 (1.10)	1.00 (0.95)	0.90

6.4.5   Integral response deformation method

The integral response deformation method is a practical calculation method for seismic analysis of underground structures proposed in recent years. At present, the integral response displacement method for seismic analysis of underground structures with complex sections has been adopted in the *Earthquake Design Standard for Underground Structures* GB/T 51336—2018. The method is to establish the soil-structure integral model to consider the interaction between the soil and the structure, and avoids the determination and use of the foundation spring in the classical response displacement method, and also make up for the limitation of the classical response displacement method which is only applicable to the seismic analysis of the underground structure with regular section shape.

In this method, the foundation spring stiffness is calculated by the finite element model, and the soil-structure interaction model is used to replace the spring-beam model in the response displacement method. In the soil model of the excavated structure, the forced displacement is applied along the periphery of the hole to calculate the equivalent load, and the structural inertia force and the shear force around the structure are calculated by the same procedure as the response displacement method.

The calculation steps by the integral response displacement method are as follows:

(1) Seismic response analysis of one-dimensional soil is carried out by using EERA (equivalent-linear earthquake site response analysis) to solve the response displacement,

shear force and acceleration of soil layer;

(2) Equivalent load for response displacement of soil layer is solved and nodal reaction force is calculated;

(3) Calculating stratum shear force and structural inertia force;

(4) Establishing the calculation model of integral response displacement method;

(5) Checking the force and deformation of structural elements;

(6) Evaluating the seismic capacity of underground structures.

参考文献

［1］中华人民共和国国家标准.《建筑抗震设计规范》GB 50011—2010（2016 年版）［S］.北京：中国建筑工业出版社，2016.

［2］中华人民共和国国家标准.《建筑结构荷载规范》GB 50009—2012［S］.北京：中国建筑工业出版社，2012.

［3］中华人民共和国国家标准.《建筑工程抗震设防分类标准》GB 50223—2008［S］.北京：中国建筑工业出版社，2008.

［4］中华人民共和国国家标准.《混凝土结构设计规范》GB 50010—2010（2015 年版）［S］.北京：中国建筑工业出版社，2015.

［5］中华人民共和国国家标准.《地下结构抗震设计标准》GB/T 51336—2018［S］.北京：中国建筑工业出版社，2019.

［6］Anil K. Chopra. Dynamics of structures：theory and applications to earthquake engineering (4th Edition)［M］. USA：Prentice Hall，2011.

［7］Shashikant K. Duggal. Earthquake resistant design of structure［M］. UK：Oxford University Press 2013.

［8］龙驭球，包世华，袁驷.结构力学（第 4 版）［M］.北京：高等教育出版社，2018.

［9］王社良.抗震结构设计（第 4 版）［M］.武汉：武汉理工大学出版社，2015.

［10］郭继武.建筑抗震设计（第 3 版）［M］.北京：中国建筑工业出版社，2011.

［11］杨德健.建筑结构抗震设计［M］.北京：人民交通出版社，2011.

［12］王显利.工程结构抗震设计［M］.北京：科学出版社，2008.

［13］周锡元，吴育才.工程结构抗震的新发展［M］.北京：清华大学出版社，2006.

［14］尚守平.结构抗震设计［M］.北京：高等教育出版社，2003.

［15］王显利.工程结构抗震设计［M］.北京：科学出版社，2008.

［16］张玉敏，苏幼坡，韩建强，等.建筑结构与抗震设计［M］.北京：清华大学出版社，2016.

［17］张耀庭，潘鹏.建筑结构抗震设计［M］.北京：机械工业出版社，2018.

［18］郑永来，杨林德，李文艺，等.地下结构抗震［M］.上海：同济大学出版社，2011.

［19］庄海洋，陈国兴.地铁地下结构抗震［M］.北京：科学出版社，2017.

参考文献